T0211105

Lecture Notes in Economics and Mathematical Systems

Volume 690

This series reports on new developments in mathematical economics, economic theory, econometrics, operations research and mathematical systems.

The series welcomes proposals for:

1. Research monographs
2. Lectures on a new field or presentations of a new angle in a classical field
3. Seminars on topics of current research
4. Reports of meetings provided they are of exceptional interest and devoted to a single topic.

In the case of a research monograph, or of seminar notes, the timeliness of a manuscript may be more important than its form, which may be preliminary or tentative.

Manuscripts should be no less than 150 and preferably no more than 500 pages in length.

The series and the volumes published in it are indexed by Scopus and ISI (selected volumes).

More information about this series at http://www.springer.com/series/300

Jaroslav Ramík

Pairwise Comparisons Method

Theory and Applications in Decision Making

 Springer

Jaroslav Ramík
School Business Administration
Silesian University in Opava
Karvina, Czech Republic

This work has been supported by the project of GACR No. 18-01246S.

ISSN 0075-8442 ISSN 2196-9957 (electronic)
Lecture Notes in Economics and Mathematical Systems
ISBN 978-3-030-39890-3 ISBN 978-3-030-39891-0 (eBook)
https://doi.org/10.1007/978-3-030-39891-0

To my grandchildren Evka and Pavlik.

Preface

A pairwise comparisons matrix is the result of pairwise comparisons a powerful method in multi-criteria optimization and decision-making. Comparing two elements is easier than comparing three and more elements at the same time. The decision-maker assigns the value representing the element of the pairwise comparisons matrix according to his/her wishes and knowledge. In this book, we shall investigate various relationships between pairwise comparisons matrices. The input data in real problems are usually not exact and can instead be characterized by uncertain values. Here we apply fuzzy sets, intuitionistic fuzzy sets as well as random value sets and, of course, the classical crisp sets. Considering matrices and vectors with uncertain coefficients values is therefore of great practical importance. The book is based on the author's 20 years of research in this area.

Český Těšín, Karvina, Czech Republic   Jaroslav Ramík
December 2019

Introduction

In various fields of evaluation, selection, and prioritization processes decision-maker(s) try to find the best alternative(s) from a feasible set of alternatives. In many cases, the comparison of different alternatives according to their desirability in decision problems cannot be done using only a single criterion or one decision-maker. In many decision-making problems, procedures have been established to combine opinions about alternatives related to different points of view. These procedures are often based on *pairwise comparisons*, in the sense that processes are linked to some degree of preference for one alternative over another. According to the nature of the information expressed by the decision-maker, for every pair of alternatives, different representation formats can be used to express preferences, e.g., multiplicative preference relations, additive preference relations, fuzzy preference relations, interval-valued preference relations, and also linguistic preference relations.

The presented monograph is devoted to pairwise comparisons. It consists of two parts: in the first part, theoretical aspects of pairwise comparisons are investigated from various points of views. The pairwise comparisons matrix (PCM), a fundamental tool for further investigation, is firstly viewed as a deterministic matrix with given elements. Then, in the following chapters, it is investigated under uncertainty, either as a matrix with vague elements (fuzzy and/or intuitionistic fuzzy ones), and also as random elements.

In the second part, theoretical results of the first part are applied in the three most popular multi-criteria decision-making methods: the Analytic Hierarchy Process (AHP), PROMETHEE, and TOPSIS. In these methods, pairwise comparisons play a leading role with a decisive impact. From this point of view, all well-known methods are reconsidered in new and broader perspectives.

In Chap. 1 basic preliminary concepts and elementary results are presented. They are used later in the following chapters: groups, alo-groups, fuzzy sets, fuzzy relations, fuzzy numbers, triangular fuzzy numbers, fuzzy matrices, and others.

Chapter 2 deals with pairwise comparisons matrices. In a multi-criteria decision-making context, a pairwise comparisons matrix is a helpful tool to determine the weighted ranking of a set of alternatives or criteria. The entry of the matrix can assume different meanings: it can be a preference ratio (multiplicative case) or a preference difference (additive case), or it belongs to the unit interval and measures the distance from the indifference that is expressed by the value 0.5 (fuzzy case). When comparing two elements, the decision-maker assigns the value from a scale to any pair of alternatives representing the element of the pairwise preference matrix. Here, we investigate the transitivity and consistency of preference matrices being understood differently with respect to the type of preference matrix. By various methods and from various types of preference matrices we obtain corresponding priority vectors for the final ranking of alternatives. The obtained results are also applied to situations where some elements of the fuzzy preference matrix are missing. Illustrative numerical examples are provided.

In Chap. 3, a unified framework for pairwise comparisons matrices based on abelian linearly ordered groups is presented. In this chapter the results from Chap. 2 are generalized for alo-groups.

Chapter 4 is focused on pairwise comparisons matrices with fuzzy and intuitionistic fuzzy elements. Fuzzy elements and intuitionistic fuzzy elements of the pairwise comparisons matrix are applied whenever the decision-maker is not sure about the value of his/her evaluation of the relative importance of the elements in question. We deal particularly with pairwise comparisons matrices with fuzzy number components and investigate some properties of such matrices. In comparison with pairwise comparison matrices with crisp components investigated in the previous chapter, here we investigate pairwise comparisons matrices with elements from alo-groups over a real interval. Such an approach allows for generalization of additive, multiplicative, and fuzzy pairwise comparisons matrices with fuzzy elements. Moreover, we deal with the problem of measuring the inconsistency of fuzzy pairwise comparisons matrices by defining corresponding inconsistency indexes. Numerical examples are presented to illustrate the concepts and derived properties.

The aim of Chap. 5 is to review the key statistical approaches to extracting the weights of objects from a PCM and, in doing so, to relate the embedded models to the traditional linear stochastic models used in the method of pairwise comparisons. The chapter is organized as follows. First, it comprises a brief account of the method of pairwise comparisons and a formal specification of a PC matrix. Statistical approaches to the analysis of judgment matrices are then introduced within the context of pairwise comparisons, with those which are distribution-based, and those based more directly on the method of pairwise comparisons.

The first five chapters of the monograph are devoted to the theory of pairwise comparisons dealing with various aspects of the uncertainty of elements of PCMs. On the other hand, the last two chapters are focused on the practical application of pairwise comparisons in two of the most popular decision-making methods: AHP and PROMETHEE. Both methods are briefly introduced and described with a special emphasis on the features and functioning of pairwise comparisons. From this point of view, both methods are revisited.

The Analytic Hierarchy Process (AHP) investigated in Chap. 6 is a theory of relative measurement on absolute scales of both quantitative and qualitative criteria based on the paired comparison judgment of knowledgeable experts. Comparisons can serve as a tool of measurement and how a valid scale of priorities can be derived from these measurements. How to measure intangibles is the main concern of the mathematics of the AHP. But it must work with tangibles as well to give back measurements where they can be used for the tangible factors in a decision with accuracy on tangibles. A fundamental problem of decision theory is how to derive weights for a set of activities according to their importance. Importance is usually judged according to several criteria. Each criterion may be shared by some or by all of the activities. The criteria may, for example, be objectives which the activities have been devised to fulfill. This is a process of multiple criteria decision-making which we study here through a theory of measurement in a hierarchical structure.

In Chap. 7, the methods of the PROMETHEE type and TOPSIS type are based on pairwise comparisons of n alternatives by m criteria. We assume, that the criteria are cardinal and their relative importance is given by the weights. We can use any evaluation system, not only multiplicative, but also additive, fuzzy, or any other one, in order to obtain the corresponding priority vector suitable for the given DM problem. The result of pairwise comparisons of alternatives is a number from the interval $[0; 1]$, and the intensity depends on the difference of values of the criterion. If the criterion is maximizing (i.e., the higher the value, the better), then the greater the difference, the higher the preference. A particular preference is given by the preference functions. The methods of the PROMETHEE type utilize six basic types of preference functions, and, moreover, each preference function has an associated threshold of preference, threshold of indifference, and standard deviation.

The TOPSIS is a multi-criteria decision method, based on the concept that the optimal alternative should have the shortest "distance" from the positive ideal solution (PIS) and the longest "distance" from the negative ideal solution (NIS). An important part of TOPSIS model is based on pairwise comparisons.

The chapter is closed with two case studies demonstrating the FuzzyDAME, a special software tool (Excell add-in) for solving MCDM problems, available for the readers free of charge on the specific website mentioned in Chap. 7.

Recently, the literature concerning pairwise comparisons has become extensive and is still rapidly growing. Among others, this monograph is based on 27 publications: six books, ([1, 4, 5, 6, 10, 26]), 13 papers published in international journals listed in Web of Science and SCOPUS ([2, 3, 8, 12, 13, 15, 16, 17, 20, 21, 22, 23, 25]), and eight papers in the proceedings of the conferences listed in WoS ([7, 9, 11, 14, 18, 19, 24, 27]). Theoretical aspects of PCMs are investigated in the journal papers, whereas the conference papers are mostly devoted to special software for solving MCDM problems. The above-mentioned publications have been created and published over the time of 20 years, from 2000 up till 2019. The main author of all the mentioned publications is Jaroslav Ramík, the author of this monograph, with contributions by colleagues from the Silesian University in Opava, Czechia, School of Business Administration in Karvina, and several other colleagues from the Czech universities.

It should be mentioned that a lot of the material and, particularly, results cited in this monograph, are new, original, still unpublished. Among others, Chap. 3,

Sect. 3.6 "Strong transitive and weak consistent PCM" is totally new and also Sect. 3.7 is a completely rewritten paper [20]. Also, Sect. 3.9 "What is the best evaluation method for pairwise comparisons" is a novel case study, still unpublished. Chapter 4 contains mostly new ideas and results, too. In particular, here, the concepts of weak and strong consistency of PCMs with fuzzy elements have been essentially changed when compared with papers [22] and [25] and some interesting new results have been obtained. The stochastic approaches to PCMs investigated in Chap. 5 are not new, however, although in this chapter they are reconsidered with respect to alternative uncertainty concepts: fuzzy and intuitionistic fuzzy elements. Chapter 6 is devoted to the most popular MCDM method—AHP, where pairwise comparisons method plays a dominant role. We have extended AHP by supplementing it with new features of PCMs, thus enabling novel branches of application.

Karvina, Czech Republic   Jaroslav Ramík
December 2019

References

1. Gavalec M, Ramik J, Zimmermann K (2014) Decision making and optimization—special matrices and their applications in economics and management. Springer International Publishing, Switzerland, Cham-Heidelberg-New York-Dordrecht-London
2. Kulakowski K, Mazurek J, Ramik J et al (2019) When is the condition of preservation met? Eur J Oper Res 277(1):248–254
3. Mazurek J, Ramik J (2019) Some new properties of inconsistent pairwise comparisons matrices. Int J Approximate Reasoning 113:119–132
4. Ramik J (1999) Analyticky hierarchicky proces (AHP). OPF Karvina publishing, Slezska univerzita, p 195
5. Ramik J (2000) Analyticky hierarchicky proces (AHP) a jeho vyuziti v malem a strednim podnikani (AHP and its application in small and medium businesses). OPF Karvina publishing, Slezska univerzita, p 217
6. Ramik J, Vlach M (2001) Generalized concavity in optimization and decision making. Kluwer Publishing Company, Boston-Dordrecht-London
7. Ramik J, Perzina R (2006) Fuzzy ANP—a new method and case study. In: Proceedings of the 24th international conference mathematical methods in economics. University of Western Bohemia
8. Ramik J (2007) A decision system using ANP and fuzzy inputs. Int J Innovative Comput Inf Control 3(4):825–837
9. Ramik J, Perzina R (2008) Microsoft excel add-in for solving multicriteria decision problems in fuzzy environment. In: Proceedings of the 26th international conference mathematical methods in economics. Technical University of Liberec
10. Ramik J, Perzina R (2008) Moderni metody hodnoceni a rozhodovani (Modern methods in evaluation and decision making). Silesian University in Opava, School of business administration publishing, Karvina, p 252
11. Ramik J, Perzina R (2008) Method for solving fuzzy MCDM problems with dependent criteria. In: Proceedings of the joint 4th international conference on soft computing and intelligent systems and 9th international symposium on advanced intelligent systems. Nagoya University, Nagoya, pp 1323–1328

12. Ramik J, Perzina R (2010) A method for solving fuzzy multicriteria decision problems with dependent criteria. Fuzzy Optimi Decis Making 9(2):123–141
13. Ramik J, Korviny P (2010) Inconsistency of pairwise comparison matrix with fuzzy elements based on geometric mean. Fuzzy Sets Syst 161:1604–1613
14. Ramik J, Perzina R (2012) DAME—Microsoft Excel add-in for solving multicriteria decision problems with scenarios. In: Proceedings of the 30th international conference mathematical methods in economics, Silesian University, School of Business Administration
15. Ramik J, Vlach M (2012) Aggregation functions and generalized convexity in fuzzy optimization and decision making. Annals Operat Res 191:261–276
16. Ramik J, Vlach M (2013) Measuring consistency and inconsistency of pair comparison systems. Kybernetika 49(3):465–486
17. Ramik J, Perzina R (2014) Solving decision problems with dependent criteria by new fuzzy multicriteria method in excel. J Bus Manage 3:1–16
18. Ramik J, Perzina R (2014) Microsoft Excel as a tool for solving multicriteria decision problems. In: Proceedings of the 18th annual conference, KES-2014 Gdynia, Poland, September, Procedia Computer Science 35, Elsevier, Gdynia, pp 1455–1463
19. Ramik J, Perzina R (2014) Solving multicriteria decision making problems using Microsoft Excel. In: Proceedings of the 32nd international conference mathematical methods in economics, September 10–12, Olomouc, Palacky University, Faculty of Science, Olomouc, pp 777–782
20. Ramik J (2014) Incomplete fuzzy preference matrix and its application to ranking of alternatives. Int J Intelligent Syst 29(8):787–806
21. Ramik J (2015) Isomorphisms between fuzzy pairwise comparison matrices. Fuzzy Optim Decis Making 14:199–209
22. Ramik J (2015) Pairwise comparison matrix with fuzzy elements on alo-groups. Inf Sci 297:236–253
23. Ramik J (2017) Ranking alternatives by pairwise comparisons matrix and priority vector. Sci Ann Econ Bus 64:85–95
24. Ramik J (2017) Strict and strong consistency in pairwise comparisons matrix with fuzzy elements. In: Novak V et al (eds) Proceedings of the 20th Czech–Japan seminar on data analysis and decision making under uncertainty. Pardubice, Czech Republic, September 17–20, University of Ostrava, pp 176–187 (2017).
25. Ramik J (2018) Strong reciprocity and strong consistency in pairwise comparison matrix with fuzzy elements. Fuzzy Optim Decis Making 17:337–355
26. Ramik J (2018) Decision analysis for managers. SBA Karvina publishing, Silesian University in Opava, p 170
27. Ramik J (2019) New approach how to generate priority vector to pairwise comparisons matrix with fuzzy elements. In: Inuiguti M et al (eds) Proceedings of the 22th Czech–Japan seminar on data analysis and decision making under uncertainty. Novy Svetlov, Czech Republic, September 25–28, Charles University Prague, 151–162

Contents

Part I
Pairwise Comparisons Method—Theory

Chapter 1
Preliminaries

1.1 Fuzzy Sets

In order to define the concept of a fuzzy subset of a given set X within the framework of standard set theory we are motivated by the concept of the upper level set of a function, see also [7]. Throughout this chapter, X is a nonempty set. All propositions are stated without proofs, but the reader can find them, e.g., in [9].

Definition 1.1 Let X be a nonempty set. A *fuzzy subset A of X* is the family of subsets $A_\alpha \subset X$, where $\alpha \in [0; 1]$, satisfying the following properties:

$$A_0 = X, \tag{1.1}$$

$$A_\beta \subset A_\alpha \qquad \text{whenever } 0 \le \alpha < \beta \le 1, \tag{1.2}$$

$$A_\beta = \bigcap_{0 \le \alpha < \beta} A_\alpha \qquad \text{for all } \beta \in [0, 1]. \tag{1.3}$$

A fuzzy subset A of X will also be called a *fuzzy set* . The class of all fuzzy subsets of X is denoted by $\mathscr{F}(X)$.

Definition 1.2 Let $A = \{A_\alpha\}_{\alpha \in [0;1]}$ be a fuzzy subset of X. The $\mu_A : X \to [0; 1]$ defined by

$$\mu_A(x) = \sup\{\alpha \mid \alpha \in [0; 1], \ x \in A_\alpha\} \tag{1.4}$$

is called the membership function of A, and the value $\mu_A(x)$ is called the membership degree of x in the fuzzy set A.

Definition 1.3 Let A be a fuzzy subset of X. The core of A, Core(A), is defined by

$$\text{Core}(A) = \{x \in X \mid \mu_A(x) = 1\}.$$

If the core of A is nonempty, then A is said to be normalized. The support of A, Supp(A), is defined by

© The Editor(s) (if applicable) and The Author(s), under exclusive
license to Springer Nature Switzerland AG 2020
J. Ramik, *Pairwise Comparisons Method*, Lecture Notes in Economics
and Mathematical Systems 690, https://doi.org/10.1007/978-3-030-39891-0_1

$$\text{Supp}(A) = \text{Cl}(\{x \in X \mid \mu_A(x) > 0\}).$$

The height of A, $\text{Hgt}(A)$, is defined by

$$\text{Hgt}(A) = \sup\{\mu_A(x) \mid x \in X\}.$$

The upper level set of the membership function μ_A of A at $\alpha \in [0; 1]$ is denoted by $[A]_\alpha$ and called the α-cut of A, that is,

$$[A]_\alpha = \{x \in X \mid \mu_A(x) \geq \alpha\}. \tag{1.5}$$

Note that if A is normalized, then $\text{Hgt}(A) = 1$, but not vice versa.

In the following two propositions, we show that the family generated by the upper level sets of a function $\mu : X \to [0; 1]$, satisfies conditions (1.1)–(1.3), thus, it generates a fuzzy subset of X and the membership function μ_A defined by (1.4) coincides with μ. Moreover, for a given fuzzy set $A = \{A_\alpha\}_{\alpha \in [0;1]}$, every α-cut $[A]_\alpha$ given by (1.5) coincides with the corresponding A_α.

Proposition 1.1 *Let $\mu : X \to [0; 1]$ be a function and let $A = \{A_\alpha\}_{\alpha \in [0,1]}$ be a family of its upper level sets, i.e., $A_\alpha = \{x \in X \mid \mu(x) \geq \alpha\}$ for all $\alpha \in [0; 1]$. Then A is a fuzzy subset of X and μ is the membership function of A.*

Proposition 1.2 *Let $A = \{A_\alpha\}_{\alpha \in [0;1]}$ be a fuzzy subset of X and let $\mu_A : X \to [0; 1]$ be the membership function of A. Then for each $\alpha \in [0; 1]$ the α-cut $[A]_\alpha$ is equal to A_α.*

These results allow for introducing a natural one-to-one correspondence between fuzzy subsets of X and real-valued functions mapping X to $[0; 1]$. Any fuzzy subset A of X is given by its membership function μ_A and vice versa, any function $\mu : X \to [0; 1]$ uniquely determines a fuzzy subset A of X, with the property that the membership function μ_A of A is μ.

The notions of inclusion and equality extend to fuzzy subsets as follows. Let $A = \{A_\alpha\}_{\alpha \in [0;1]}$, $B = \{B_\alpha\}_{\alpha \in [0;1]}$ be fuzzy subsets of X. Then

$$A \subset B \text{ if } A_\alpha \subset B_\alpha \quad \text{for each } \alpha \in [0; 1], \tag{1.6}$$

$$A = B \text{ if } A_\alpha = B_\alpha \quad \text{for each } \alpha \in [0; 1]. \tag{1.7}$$

Proposition 1.3 *Let $A = \{A_\alpha\}_{\alpha \in [0;1]}$ and $B = \{B_\alpha\}_{\alpha \in [0;1]}$ be fuzzy subsets of X. Then the following holds*

$$A \subset B \text{ if and only if } \mu_A(x) \leq \mu_B(x) \quad \text{for all } x \in X, \tag{1.8}$$

$$A = B \text{ if and only if } \mu_A(x) = \mu_B(x) \quad \text{for all } x \in X. \tag{1.9}$$

A subset of X can be considered as a special fuzzy subset of X where all its members defining a family consist of the same elements. This is formalized in the following definition.

Definition 1.4 Let A be a subset of X. The fuzzy subset $\{A_\alpha\}_{\alpha \in [0;1]}$ of X defined by $A_\alpha = A$ for all $\alpha \in]0; 1]$ is called a *crisp fuzzy subset of X generated by A*. A fuzzy subset of X generated by some $A \subset X$ is called a *crisp fuzzy subset of X* or briefly a *crisp subset of X*.

Proposition 1.4 *Let $\{A\}_{\alpha \in [0;1]}$ be a crisp subset of X generated by A. Then the membership function of $\{A\}_{\alpha \in [0;1]}$ is equal to the characteristic function of A.*

By Definition 1.4, the set $\mathscr{P}(X)$ of all subsets of X can naturally be embedded into the set of all fuzzy subsets of X and we can write $A = \{A_\alpha\}_{\alpha \in [0;1]}$ if $\{A_\alpha\}_{\alpha \in [0;1]}$ is generated by $A \subset X$. According to Proposition 1.4, we have in this case $\mu_A = \chi_A$. In particular, if A contains only one element a of X, that is, $A = \{a\}$, then we write $a \in \mathscr{F}(X)$ instead of $\{a\} \in \mathscr{F}(X)$ and χ_a instead of $\chi_{\{a\}}$.

1.2 Extension Principle

The purpose of the following definition called the *extension principle* (proposed by Zadeh in [12, 13]) is to extend functions or operations having crisp arguments to functions or operations with fuzzy set arguments. Zadeh's methodology can be cast in a more general setting of carrying a membership function via a mapping, see e.g., [6]. There exist other generalizations for set-to-set mappings; see e.g., [6, 10]. From now on, X and Y are nonempty sets.

Definition 1.5 *(Extension Principle)* Let X, Y be sets, $f : X \to Y$ be a mapping. The mapping $\tilde{f} : \mathscr{F}(X) \to \mathscr{F}(Y)$ defined for all $A \in \mathscr{F}(X)$ with $\mu_A : X \to [0, 1]$ and all $y \in Y$ by

$$\mu_{\tilde{f}(A)}(y) = \begin{cases} \sup\{\mu_A(x) \mid x \in X, \ f(x) = y\} & \text{if } f^{-1}(y) \neq \emptyset, \\ 0 & \text{otherwise,} \end{cases} \tag{1.10}$$

is called a *fuzzy extension of f*.

By formula (1.10) we define the membership function of the image of the fuzzy set A by fuzzy extension $\tilde{f}$. A justification of this concept is given in the following theorem stating that the mapping $\tilde{f}$ is a true extension of the mapping f when considering the natural embedding of $\mathscr{P}(X)$ into $\mathscr{F}(X)$ and $\mathscr{P}(Y)$ into $\mathscr{F}(Y)$, see [11].

Proposition 1.5 *Let X, Y be sets, $f : X \to Y$ be a mapping, $x_0 \in X$, $y_0 = f(x_0)$. If $\tilde{f} : \mathscr{F}(X) \to \mathscr{F}(Y)$ is defined by (1.10), then*

$$\tilde{f}(x_0) = y_0,$$

and the membership function $\mu_{\tilde{f}(x_0)}$ of the fuzzy set $\tilde{f}(x_0)$ is a characteristic function of y_0, i.e.,

$$\mu_{\tilde{f}(x_0)} = \chi_{y_0}. \tag{1.11}$$

A more general form of Proposition 1.5 says that the image of a crisp set by a fuzzy extension of a function is again crisp.

Proposition 1.6 *Let X, Y be sets, $f : X \to Y$ be a mapping, $A \subset X$. Then*

$$\tilde{f}(A) = f(A)$$

and the membership function $\mu_{\tilde{f}(A)}$ of $\tilde{f}(A)$ is a characteristic function of the set $f(A)$, i.e.,

$$\mu_{\tilde{f}(A)} = \chi_{f(A)}. \tag{1.12}$$

In the following sections the extension principle will be used in different settings for various sets X and Y, and also for different classes of mappings and relations, [1, 8].

1.3 Binary Relations, Valued Relations, and Fuzzy Relations

In the classical set theory, a *binary relation* R between the elements of sets X and Y is defined as a subset of the Cartesian product $X \times Y$, that is, $R \subset X \times Y$. A valued relation on $X \times Y$ will be a fuzzy subset of $X \times Y$.

Definition 1.6 A *valued relation R on $X \times Y$* is a fuzzy subset of $X \times Y$. The set of all valued relations on $X \times Y$ is denoted by $\mathscr{F}(X \times Y)$.

The valued relations are sometimes called fuzzy relations, however, we reserve this name for valued relations defined on $\mathscr{F}(X) \times \mathscr{F}(Y)$, which will be defined later.

Every binary relation R, where $R \subset X \times Y$, is embedded into the class of valued relations on $X \times Y$ by its characteristic function χ_R being understood as its membership function μ_R. In this sense, any binary relation is valued.

In particular, any function $f : X \to Y$ is considered as a binary relation, that is, as a subset R_f of $X \times Y$, where

$$R_f = \{(x, y) \in X \times Y \mid y = f(x)\}. \tag{1.13}$$

Here, R_f may be identified with the valued relation by its characteristic function

$$\mu_{R_f}(x, y) = \chi_{R_f}(x, y) \tag{1.14}$$

for all $(x, y) \in X \times Y$, where

$$\chi_{R_f}(x, y) = \chi_{f(x)}(y). \tag{1.15}$$

In particular, if $Y = X$, then each valued relation R on $X \times X$ is a fuzzy subset of $X \times X$, and it is called a valued relation on X instead of on $X \times X$.

Definition 1.7 A valued relation R on X is

(i) *reflexive* if for each $x \in X$
$$\mu_R(x, x) = 1;$$

(ii) *symmetric* if for each $x, y \in X$

$$\mu_R(x, y) = \mu_R(y, x);$$

(iii) *min-transitive* if for each $x, y, z \in X$

$$\min\{\mu_R(x, y), \mu_R(y, z)\} \le \mu_R(x, z);$$

(iv) *separable* if
$$\mu_R(x, y) = 1 \text{ if and only if } x = y;$$

(v) *equivalence* if R is reflexive, symmetric, and min-transitive.

Definition 1.8 Let R be a valued relation on $X \times Y$ and let $N : [0; 1] \to [0; 1]$ be a negation, i.e., $N(x) = 1 - x$ for all $x \in [0; 1]$.

(i) A valued relation R^{-1} on $Y \times X$ is the inverse of R if $\mu_{R^{-1}}(y, x) = \mu_R(x, y)$ for each $x \in X$ and $y \in Y$.
(ii) A valued relation $\mathscr{C}_N R$ on $X \times Y$ is the complement of R if $\mu_{\mathscr{C}_N R}(x, y) = N(\mu_R(x, y))$ for each $x \in X$ and $y \in Y$. If N is the standard negation, then the index N is omitted.
(iii) If μ_R is upper semicontinuous on $X \times Y$, then R is called closed.

Let X, Y be nonempty sets. Consider a valued relation R on $X \times Y$ given by the membership function $\mu_R : X \times Y \to [0; 1]$. In order to extend this function with crisp arguments to function with fuzzy arguments, we apply the extension principle (3.72) in Definition 1.5. Then we obtain a mapping $\tilde{\mu}_R : \mathscr{F}(X \times Y) \to \mathscr{F}([0; 1])$, that is, values of $\tilde{\mu}_R$ are fuzzy subsets of $[0; 1]$.

Definition 1.9 A fuzzy subset of $\mathscr{F}(X) \times \mathscr{F}(Y)$ is called a *fuzzy relation on* $\mathscr{F}(X) \times \mathscr{F}(Y)$. The set of all fuzzy relations on $\mathscr{F}(X) \times \mathscr{F}(Y)$ is denoted by $\mathscr{F}(\mathscr{F}(X) \times \mathscr{F}(Y))$.

1.4 Fuzzy Quantities, Fuzzy Numbers, and Fuzzy Intervals

In this section, we are concerned with fuzzy subsets of the real line. Therefore we have $X = \mathbf{R}$ and $\mathscr{F}(X) = \mathscr{F}(\mathbf{R})$.

Definition 1.10 (i) A fuzzy subset $A = \{A_\alpha\}_{\alpha \in [0; 1]}$ of $\mathbf{R}$ is called a *fuzzy quantity*. The set of all fuzzy quantities will be denoted by $\mathscr{F}(\mathbf{R})$.

(ii) A fuzzy quantity $A = \{A_\alpha\}_{\alpha \in [0;1]}$ is called a *fuzzy interval* if A_α is nonempty, convex, and closed subset of $\mathbf{R}$ for all $\alpha \in [0; 1]$. The set of all fuzzy intervals will be denoted by $\mathscr{F}_I(\mathbf{R})$.

(iii) A fuzzy interval A is called a *fuzzy number* if its core is a singleton. The set of all fuzzy numbers will be denoted by $\mathscr{F}_N(\mathbf{R})$.

Notice that the membership function $\mu_A : \mathbf{R} \to [0; 1]$ of a fuzzy interval A is quasi-concave on $\mathbf{R}$, that is, for all $x, y \in \mathbf{R}$, $x \neq y$, $\lambda \in]0; 1[$, the following inequality holds

$$\mu_A(\lambda x + (1 - \lambda)y) \geq \min\{\mu_A(x), \mu_A(y)\}.$$

By Definition 1.10, each fuzzy interval is normalized, since $\mathrm{Core}(A) = [A]_1$ is nonempty, that is, there exists an element $x_0 \in \mathbf{R}$ with $\mu_A(x_0) = 1$. Then $\mathrm{Hgt}(A) = 1$. Moreover, the restriction of the membership function μ_A to $]-\infty; x_0]$ is non-decreasing and the restriction of μ_A to $[x_0; +\infty[$ is a non-increasing function.

A fuzzy interval A has an upper semicontinuous membership function μ_A or, equivalently, for each $\alpha \in]0; 1]$ the α-cut $[A]_\alpha$ is a closed subinterval in $\mathbf{R}$. Such a membership function μ_A, and the corresponding fuzzy interval A, can be fully described by a quadruple (l, r, F, G), where $l, r, \in \mathbf{R}$ with $l \leq r$, and F, G are non-increasing left continuous functions mapping $]0; +\infty[$ into $[0; 1[$, by setting

$$\mu_A(x) = \begin{cases} F(l - x) & \text{if } x \in]-\infty; l[, \\ 1 & \text{if } x \in [l; r], \\ G(x - r) & \text{if } x \in]r; +\infty[. \end{cases} \tag{1.16}$$

We shall briefly write $A = (l, r, F, G)$. As the ranges of F and G are included in $[0; 1[$, we have $\mathrm{Core}(A) = [l; r]$. We can see that the functions F, G describe the left and right "shape" of μ_A, respectively. Observe also that each crisp number $x_0 \in \mathbf{R}$ and each crisp interval $[a; b] \subset \mathbf{R}$ belongs to $\mathscr{F}_I(\mathbf{R})$, as they may be equivalently expressed by the characteristic functions $\chi_{\{x_0\}}$ and $\chi_{[a;b]}$, respectively. These characteristic functions can be also described in the form (1.16) with $F(x) = G(x) = 0$ for all $x \in]0; +\infty[$.

Example 1.1 (Gaussian fuzzy number) Let $a \in \mathbf{R}$, $\gamma \in]0; +\infty[$, and let $A = (a, a, G, G)$ where

$$G(x) = e^{-\frac{x^2}{\gamma}}.$$

Then the membership function μ_A of A is given by

$$\mu_A(x) = G(x - a) = e^{-\frac{(x-a)^2}{\gamma}}.$$

A class of more specific fuzzy intervals of $\mathscr{F}_I(\mathbf{R})$ is obtained, if the α-cuts are required to be bounded intervals. Let $l, r, \in \mathbf{R}$ with $l \leq r$, let $\gamma, \delta \in [0; +\infty[$ and let L, R be continuous and strictly decreasing functions mapping interval $[0; 1]$ into

$[0; +\infty[$, i.e., $L, R : [0; 1] \to [0; +\infty[$. Moreover, assume that $L(1) = R(1) = 0$, and for each $x \in \mathbf{R}$ let

$$
\mu_A(x) = \begin{cases}
L^{(-1)}\left(\frac{l-x}{\gamma}\right) & \text{if } x \in]l - \gamma; l[, \gamma > 0, \\
1 & \text{if } x \in [l; r], \\
R^{(-1)}\left(\frac{x-r}{\delta}\right) & \text{if } x \in]r; r + \delta[, \delta > 0, \\
0 & \text{otherwise,}
\end{cases}
$$

where L^{-1}, R^{-1} are inverse functions of L, R, respectively. We shall write $A = (l, r, \gamma, \delta)_{LR}$, and say that A is an (L, R)-*fuzzy interval*. The set of all (L, R)-fuzzy intervals will be denoted by $\mathscr{F}_{LRI}(\mathbf{R})$. The values of γ, δ are called the *left* and the *right spread of A*, respectively. Observe that $\text{Supp}(A) = [l - \gamma; r + \delta]$, $\text{Core}(A) = [l; r]$ and $[A]_\alpha$ is a compact interval for every $\alpha \in]0; 1]$. If $r = l$, then A is an (L, R)-*fuzzy number*. The set of all (L, R)-fuzzy numbers will be denoted by $\mathscr{F}_{LRN}(\mathbf{R})$.

Particularly important fuzzy intervals are so called *trapezoidal fuzzy intervals*, where $L(x) = R(x) = 1 - x$ for all $x \in [0; 1]$. In this case, the subscript LR will be omitted in the notation. If $l = r$, then $A = (r, r, \gamma, \delta)$ is called a *triangular fuzzy number* and the notation is simplified to: $A = (r, \gamma, \delta)$, or, equivalently, $A = (a^L, a^M, a^R)$, see below. In the following section we shall introduce the concept of matrices with fuzzy elements, particularly, with (L, R)-fuzzy numbers.

1.5 Matrices with Fuzzy Elements

From now on, we shall denote any fuzzy set A by the symbol with tilde, i.e., $\tilde{A}$. An $m \times n$ matrix $\tilde{A} = \{\tilde{a}_{ij}\}$, where $\tilde{a}_{ij}$, $i \in \{1, \cdots, m\}$, $j \in \{1, \cdots, n\}$ are fuzzy quantities, is called a *matrix with fuzzy elements*, i.e.,

$$
\tilde{A} = \begin{pmatrix}
\tilde{a}_{11} & \tilde{a}_{12} & \cdots & \tilde{a}_{1n} \\
\tilde{a}_{21} & \tilde{a}_{22} & \cdots & \tilde{a}_{2n} \\
\vdots & \vdots & \ddots & \vdots \\
\tilde{a}_{m1} & \tilde{a}_{m2} & \cdots & \tilde{a}_{mn}
\end{pmatrix}. \tag{1.17}
$$

In practice, triangular fuzzy numbers introduced in Chap. 4 are suitable for modeling fuzzy quantities by DMs. A *triangular fuzzy number* $\tilde{a} \in \mathscr{F}_{LRN}(\mathbf{R})$ can be equivalently expressed by a triple of real numbers, i.e., $\tilde{a} = (a^L, a^M, a^U)$, where a^L is the *Lower number*, a^M is the *Middle number*, and a^U is the *Upper number*, $a^L \leq a^M \leq a^U$. If $a^L = a^M = a^U$, then $\tilde{a}$ is the crisp number (non-fuzzy number). Evidently, the set of all crisp numbers is isomorphic to the set of real numbers. If $a^L \neq a^M \neq a^U$, then the *membership function* $\mu_{\tilde{a}}$ of $\tilde{a}$ is assumed to be continuous, strictly increasing in the interval $[a^L; a^M]$ and strictly decreasing in $[a^M; a^U]$. Moreover, the *membership grade* $\mu_{\tilde{a}}(x)$ is equal to zero for $x \notin [a^L; a^U]$ and equal to

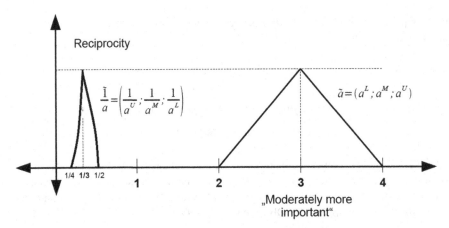

Fig. 1.1 Membership functions of $\tilde{a}$ and $\frac{\tilde{1}}{a}$

one for $x = a^M$. As usual, the membership function $\mu_{\tilde{a}}$ is assumed to be piecewise linear, see Fig. 1.1. If $a^L = a^M$ and/or $a^M = a^U$, then the membership function $\mu_{\tilde{a}}$ is discontinuous. The triangular fuzzy numbers $\tilde{a} = (a_{ij}^L, a_{ij}^M, a_{ij}^U)$ is *fuzzy positive*, if $a_{ij}^L > 0$.

The arithmetic operations $+$, $-$, $\cdot$ and $/$ can be extended to fuzzy numbers by the Extension principle, see Sect. 1.5, and also, e.g., [6].

Arithmetic operation with fuzzy numbers are defined as follows, see [6], or [9]. Let $\tilde{a} = (a^L, a^M, a^U)$ and $\tilde{b} = (b^L, b^M, b^U)$, where $a^L > 0$, $b^L > 0$, be positive triangular fuzzy numbers.

Addition: $\tilde{a} \tilde{+} \tilde{b} = (a^L + b^L, a^M + b^M, a^U + b^U)$,
Subtraction: $\tilde{a} \tilde{-} \tilde{b} = (a^L - b^U, a^M - b^M, a^U - b^L)$,
Multiplication: $\tilde{a} \tilde{\cdot} \tilde{b} = (a^L \cdot b^L; a^M \cdot b^M; a^U \cdot b^U)$,
Division: $\tilde{a} \tilde{/} \tilde{b} = (a^L/b^U, a^M/b^M, a^U/b^L)$.
In particular: $\frac{\tilde{1}}{a} = \left(\frac{\tilde{1}}{a^U}, \frac{\tilde{1}}{a^M}, \frac{\tilde{1}}{a^L} \right)$.

For using matrices with triangular fuzzy elements there exist at least following reasons:

- The membership functions of triangular fuzzy elements are usually piecewise linear, i.e., easy to understand.
- Triangular fuzzy numbers can be easily manipulated, e.g., added, multiplied.
- Crisp (non-fuzzy) numbers are special cases of triangular fuzzy numbers.
- The reciprocal matrix with triangular fuzzy elements can be considered by the DM as a model for his/her fuzzy pairwise preference representations concerning n elements (e.g., alternatives). In this model, it is assumed that only $n(n-1)/2$ judgments are needed, and the rest are given by the reciprocity condition.
- In practice, when interval-valued matrices are employed, the DM often gives ranges narrower than his or her actual perception would authorize, because he/she might

be afraid of expressing information which is too imprecise. On the other hand, triangular fuzzy numbers express rich information because the DM provides both the support set of the fuzzy number as the range that the DM believes to surely contain the unknown ratio of relative importance, and the grades of the possibility of occurrence (i.e., membership function) within this range.

- Triangular fuzzy numbers are appropriate in group decision-making, where a^L can be interpreted as the minimum possible value of DMs judgments, a^U is interpreted as the maximum possible value of DMs judgments, and a^M—the geometric mean of the DMs judgments is interpreted as the mean value, or, the most possible value of DMs judgments, see [9].

1.6 Abelian Linearly Ordered Groups

In this section, we recall some notions and properties related to abelian linearly ordered groups. The matter of this section is based on [2–5].

Definition 1.11 An *abelian group* is a set, G, together with an operation $\odot$ (read: operation *odot*) that combines any two elements $a, b \in G$ to form another element denoted $a \odot b$. The symbol $\odot$ is a general placeholder for a concretely given operation. The set and operation, $(G, \odot)$, satisfies the following requirements known as the *abelian group axioms*:

- If $a, b \in G$, then $a \odot b \in G$. (*closure axiom.*)
- If $a, b, c \in G$, then $(a \odot b) \odot c = a \odot (b \odot c)$ holds. (*associativity axiom.*)
- There exists an element $e \in G$ called the *identity element*, such that for all elements $a \in G$, the equation $e \odot a = a \odot e = a$ holds. (*identity element axiom.*)
- For each $a \in G$, there exists an element $a^{(-1)} \in G$ called the *inverse element to* a such that $a \odot a^{(-1)} = a^{(-1)} \odot a = e$, where e is the identity element. (*inverse element axiom.*)
- For all $a, b \in G$, $a \odot b = b \odot a$. (*commutativity axiom.*)

The *inverse operation* $\div$ to $\odot$ is defined for all $a, b \in G$ as follows:

$$a \div b = a \odot b^{(-1)}.$$

In other words, an abelian group is a commutative group. A group in which the group operation is not commutative is called a "non-abelian group" or "non-commutative group".

Definition 1.12 A nonempty set G is *linearly (totally) ordered* under the order relation $\leq$, if the following statements hold for all a, b and c in G:

- *Antisymmetry*:
  If $a \leq b$ and $b \leq a$ then $a = b$;
- *Transitivity*:
  If $a \leq b$ and $b \leq c$ then $a \leq c$;

- *Totality (linearity)*:
  $a \leq b$ or $b \leq a$.
  The *strict order* relation $<$ is defined for $a, b \in G$ as

$$a < b \text{ if } a \leq b \text{ and } a \neq b.$$

Antisymmetry eliminates cases when both a precedes b and b precedes a, and $a \neq b$. A relation having the property of "totality" means that any pair of elements in the set of the relation are comparable under the relation. This also means that the set can be diagrammed as a line of elements, giving it the name "linear". Totality also implies *reflexivity*, i.e., $a \leq a$. Therefore, a total order is also a *partial order* . The partial order has a weaker form of the third condition (it only requires reflexivity, not totality). An extension of a given partial order to a total order is called a linear extension of that partial order.

Definition 1.13 Let $(G, \odot)$ be an abelian group, G be linearly ordered under $\leq$. $(G, \odot, \leq)$ is said to be an *abelian linearly ordered group*, *alo-group* for short, if for all $c \in G$

$$a \leq b \text{ implies } a \odot c \leq b \odot c. \tag{1.18}$$

It is easy to show that for $a \in G$

$$a < e \text{ if and only if } a^{-1} > e, \tag{1.19}$$

$$a > e \text{ if and only if } a^{-1} < e, \tag{1.20}$$

$$a \odot a > a \text{ for all } a > e, \tag{1.21}$$

$$a \odot a < a \text{ for all } a < e. \tag{1.22}$$

If $\mathscr{G} = (G, \odot, \leq)$ is an alo-group, then G is naturally equipped with the order topology induced by $\leq$ and $G \times G$ is equipped with the related product topology. We say that $\mathscr{G}$ is a *continuous alo-group* if $\odot$ is continuous on $G \times G$.

By Definition 1.12, an alo-group $\mathscr{G}$ is a lattice ordered group, see [3]. Hence, there exists $\max\{a, b\}$, for each pair $(a, b) \in G \times G$. Nevertheless, by (1.21), (1.22), a nontrivial alo-group $\mathscr{G} = (G, \odot, \leq)$ has neither the greatest element nor the least element.

Because of the associative property, the operation $\odot$ can be extended by induction to n-ary operation, $n > 2$, by setting

$$\bigodot_{i=1}^{n} a_i = \left(\bigodot_{i=1}^{n-1} a_i \right) \odot a_n. \tag{1.23}$$

Then, for a positive integer n, the (n)-*power* $a^{(n)}$ of $a \in G$ is defined by

$$a^{(1)} = a,$$
$$a^{(n)} = \odot_{i=1}^{n} a_i, \, a_i = a \text{ for all } i \in \{1, \cdots, n\}, n \geq 2,$$

and verifies the following properties for $a, b \in G, n \geq 2$:

$$a < b \text{ if and only if } a^{(n)} < b^{(n)} , \tag{1.24}$$

$$a^{(n)} > a \text{ for all } a > e , \tag{1.25}$$

$$a^{(n)} < a \text{ for all } a < e . \tag{1.26}$$

We can extend the meaning of power $a^{(s)}$ to the case that s is a negative integer by setting

$$a^{(0)} = e \text{ and } a^{(-n)} = \left(a^{(n)} \right)^{(-1)} = \left(a^{(-1)} \right)^{(n)} . \tag{1.27}$$

We obtain also

$$(a \div b)^{(n)} = a^{(n)} \div b^{(n)} . \tag{1.28}$$

An *isomorphism* between two alo-groups $\mathcal{G} = (G, \odot, \leq)$ and $\mathcal{G}' = (G', \circ, \preceq)$ is a bijection $h : G \to G'$ that is both a lattice isomorphism and a group isomorphism, i.e.,

$$a < b \text{ if and only if } h(a) \preceq h(b) \text{ and } h(a \odot b) = h(a) \circ h(b). \tag{1.29}$$

By the associativity of the operations $\odot$ and $\circ$, the equality in (1.29) can be extended by induction to the n-operation.

Definition 1.14 Let $\mathcal{G} = (G, \odot, \leq)$ be an alo-group. Then $\mathcal{G}$ is *divisible* if for each positive integer n and each $a \in G$ there exists the (n)-th root of a denoted by $a^{(1/n)}$, i.e., $\left(a^{(1/n)} \right)^{(n)} = a$.

The (n)-th root verifies the following property:

$$\text{if } a < b \text{ then } a^{(1/n)} < b^{(1/n)} . \tag{1.30}$$

Definition 1.15 Let $\mathcal{G} = (G, \odot, \leq)$ be a divisible alo-group, n is a positive integer. Then the $\odot$-*mean* $m_\odot(a_1, a_2, \cdots, a_n)$ *of elements* $a_1, a_2, \cdots, a_n \in G$ is defined as

$$m_\odot(a_1, a_2, \cdots, a_n) = \begin{cases} a_1 & \text{for } n = 1, \\ \left(\odot_{i=1}^{n} a_i \right)^{(1/n)} & \text{for } n > 1. \end{cases}$$

Definition 1.16 Let $\mathcal{G} = (G, \odot, \leq)$ be an alo-group. Then the function $\|.\| : G \to G$ defined by $\|a\| = \max\{a, a^{(-1)}\}$ for each $a \in G$ is called a $\mathcal{G}$-*norm*.

The $\mathscr{G}$-norm satisfies the following evident properties, $a, b \in G$, e is the identity element:

(i) $\|a\| = \|a^{(-1)}\|$;
(ii) $a \leq \|a\|$;
(iii) $\|a\| \geq e$;
(iv) $\|a\| = e$ if and only if $a = e$;
(v) $\|a^{(n)}\| = \|a\|^{(n)}$;
(vi) $\|a \odot b\| \leq \|a\| \odot \|b\|$ (*triangle inequality*).

Definition 1.17 Let $\mathscr{G} = (G, \odot, \leq)$ be an alo-group. Then the operation $d : G \times G \to G$ defined by $d(a, b) = \|a \div b\|$ for all $a, b \in G$ is called a $\mathscr{G}$-*distance*.

The $\mathscr{G}$-distance d satisfies the following evident properties; $a, b, c \in G$:

(i) $d(a, b) \geq e$;
(ii) $d(a, b) = e$ if and only if $a = b$;
(iii) $d(a, b) = d(b, a)$;
(iv) $d(a, b) \leq d(a, c) \odot d(c, b)$ (*triangle inequality*).

The proof of the following proposition is a straightforward application of the definitions.

Proposition 1.7 *Let* $\mathscr{G} = (G, \odot, \leq)$ *and* $\mathscr{G}' = (G', \circ, \preceq)$ *be alo-groups and* $h : G \to G'$ *be an isomorphism between* $\mathscr{G}$ *and* $\mathscr{G}'$, $a, b \in G$, $a', b' \in G'$. *Then*

$$d_{\mathscr{G}'}(a', b') = h(d_{\mathscr{G}}(h^{-1}(a'), h^{-1}(b'))), \tag{1.31}$$

$$d_{\mathscr{G}}(a, b) = h^{-1}(d_{\mathscr{G}'}(h(a), h(b))) \tag{1.32}$$

More definitions and properties of alo-groups of the real line **R** will be presented in Chaps. 3 and 4.

References

1. Bellman RE, Zadeh LA (1970) Decision-making in a fuzzy environment. Manag Sci 17(4):141–164
2. Bourbaki N (1990) Algebra II. Springer, Berlin
3. Cavallo B, DApuzzo L (2009) A general unified framework for pairwise comparison matrices in multicriteria methods. Int J Intell Syst 24(4):377–398
4. Cavallo B, DApuzzo L, Squillante M (2012) About a consistency index for pairwise comparison matrices over a divisible Alo-Group. Int J Intell Syst 27:153–175
5. Cavallo B, DApuzzo L (2012) Deriving weights from a pairwise comparison matrix over an alo-group. Soft Comput 16:353–366
6. Dubois D et al (2000) Fuzzy interval analysis. In: Fundamentals of fuzzy sets, series on fuzzy sets, vol 1. Kluwer Academic Publishers, Dordrecht

7. Ralescu D (1979) A survey of the representation of fuzzy concepts and its applications. In: Gupta MM, Regade RK, Yager R (eds) Advances in fuzzy sets theory and applications. North Holland, Amsterdam, pp 77–91
8. Ramík J, Římánek J (1985) Inequality relation between fuzzy numbers and its use in fuzzy optimization. Fuzzy Sets Syst 16:123–138
9. Ramík J, Vlach M (2001) Generalized concavity in optimization and decision making. Kluwer Publishers Company, Dordrecht
10. Ramík J (1986) Extension principle in fuzzy optimization. Fuzzy Sets Syst 19:29–37
11. Yager RR (1980) On a general class of fuzzy connectives. Fuzzy Sets Syst 4:235–242
12. Zadeh LA (1965) Fuzzy sets. Inform Control 8:338–353
13. Zadeh LA (1975) The concept of a linguistic variable and its application to approximate reasoning. Inf Sci Part I 8:199–249; Part II 8:301–357; Part III 9:43–80

Chapter 2
Pairwise Comparison Matrices in Decision-Making

2.1 Historical Remarks

The ability to compare things has accompanied man for the whole history of mankind. The first use of pair comparison as a formal basis for the decision procedure is attributed to the eighteenth century scholar Ramon Llull who proposed binary electoral system [24]. His method was forgotten over time but rediscovered in a similar form by Condorcet [25]. Although both Llull and Condorcet treated comparisons as binary, i.e., the result of comparisons can be either "win" or "lose" (for a given alternative). Thurstone in [103] proposed the use of pairwise comparisons in a more generalized, quantitative way. Since then the result of the single pairwise comparison can be identified with a real positive number where the values greater than 1 mean the degree to which the first alternative won, and likewise, values smaller than 1 mean the degree to which the second alternative won. Llulls electoral system was in some form reinvented by [25].

Pairwise comparisons is any process of comparing entities in pairs to judge which of the entities is preferred, or has a greater amount of some quantitative property, or, whether or not the two entities are identical or indifferent. The *method of pairwise comparisons* is used in the scientific study of preferences, attitudes, voting systems, social choice, public choice, requirements engineering, and multi-agent artificial intelligence (AI) systems. In psychology literature, it is often referred to as *paired comparison*. In various fields of evaluation, selection, and prioritization processes decision-maker(s) (DM) try to find the best alternative(s) from a feasible set of alternatives. In many cases, comparison of different alternatives according to their desirability in decision problems cannot be done using only a single criterion or one person. In many decision-making problems, procedures have been established to combine opinions about alternatives related to different points of view. These procedures are often based on pairwise comparisons, in the sense that the processes are linked to some degree of preference of one alternative over another. According to the nature of the information expressed by the DM, for every pair of alternatives different representation formats can be used to express preferences, e.g., multiplicative

J. Ramik, *Pairwise Comparisons Method*, Lecture Notes in Economics and Mathematical Systems 690, https://doi.org/10.1007/978-3-030-39891-0_2

preference relations, [52], additive preference relations, [6], fuzzy preference relations, see [31, 51, 71], interval-valued preference relations, [105], and also linguistic preference relations, [4].

Decision-making models have been studied for a long time now by many authors. A detailed bibliography can be found, for example, in the website of the International Society on Multiple Criteria Decision Making, [57]. Many well-known methods have been reviewed in [43]: Preference Modeling, Conjoint Measurement, Multi-Attribute Utility Theory (MAUT), Outranking Methods, ELECTRE, PROMETHEE, MACBETH, Fuzzy Multiple Criterion Decision Aid, Multi-objective Programming, Analytic Hierarchy/Network Process (AHP/ANP), and many others. In the former Czechoslovakia, the interest in multiple criteria decision-making (MCDM) methods has been initiated by the works of Bernard Roy in sixties and seventies of the twentieth century, see e.g., [88, 89].

2.2   State of the Art

Usually, experts are characterized by their own personal background and experience of the problem to be solved. Expert opinions may differ substantially, and some of them would not be able to efficiently express a degree of preference between two or more of the available options. This may be true when an expert does not possess a precise or sufficient level of knowledge of part of the problem, or because these experts are unable to discriminate the degree to which some options are better than others. In these situations such an expert will provide an incomplete preference matrix, see [4, 63, 105].

The usual procedures for DM problems correct this lack of knowledge of a particular expert using the information provided by the rest of the experts together with aggregation procedures [92]. Estimation of missing values in an expert incomplete preference matrix is done using only the preference values provided by these particular experts. By doing this, we assume that the reconstruction of the incomplete preference matrix is compatible with the rest of the information provided by the experts. In this chapter, we summarize various approaches to the incomplete preference matrix based on different types of degree of preference of one alternative over another, e.g., multiplicative preference relations, additive preference relations, and fuzzy preference relations [4, 6, 32, 54, 63, 71, 81].

The chapter is organized as follows. Definition of the decision-making problem is formulated and multiplicative preference relations and their properties are introduced in this and the following section. Then, the most popular methods for deriving priorities from multiplicative pairwise comparisons matrices are presented. An alternative approach for deriving the priority vector based on a special optimization problem and desirable properties of the priority vector is presented. The following sections deal with additive pairwise comparison matrices and fuzzy PCMs and some methods for deriving priorities from additive pairwise comparison matrices are discussed. In the last section some concluding considerations and remarks are presented.

2.3 Problem Definition

The *decision-making problem* (DM problem) we shall thoroughly investigate in this book and can be formulated as follows. Let $\mathscr{C} = \{c_1, c_2, ..., c_n\}$ be a finite set of alternatives ($n > 2$). These alternatives have to be ranked from best to worst, using preference information given by the decision-maker (DM) in the form of an $n \times n$ *pairwise comparisons matrix (PC matrix or PCM)* $A = \{a_{ij}\}$, where a_{ij} are elements (components, entries) of the matrix A taken from a scale S, being a subset of the set of real numbers $\mathbf{R}$.

Usually, an ordinal *ranking* of alternatives is required to obtain the best alternative(s). However, it often occurs that the ordinal ranking among alternatives is not a sufficient result and a cardinal ranking called here the *rating* is required.

Some well-known limits to our capacity to handle several alternatives at a time (so-called "cognitive overload," see [93]), make it impossible to obtain the rating directly by a priority weighting vector directly, for instance asking the DM to provide the utility values for the alternatives. Therefore, it is more suitable, and also easier, to ask the DM for his opinion over the pairs of alternatives and then, once all the necessary information over the pairs is acquired, to derive the rating for the alternatives. The most popular way of eliciting the experts' preferences by pairwise comparisons between the given alternatives is the *pairwise comparisons matrix*—a mathematical tool associated with the more general concept: *preference relation*.

Since the pairwise comparison method was proposed by L. Thurstone in the Psychological Review in 1927 [103], pairwise comparison matrices have become widely used in many well-known decision-making approaches, such as the analytic hierarchy process (AHP), [93], PROMETHEE method (see Chaps. 6 and 7 or [43]), and many others. A large number of methods deriving a ranking/rating of the alternatives have been proposed in the framework of pairwise comparisons matrices in the literature. Two well-known examples are the *eigenvector method (EVM)* in AHP, [93, 94], and the *geometric mean method (GMM)*, being in fact the *logarithmic least squares method (LLSM)* (see Sect. 2.5.4 or [34]).

In one of the most popular MCDM methods—the abovementioned analytic hierarchy process (AHP), [92], the decision problem is structured hierarchically at different levels, each level consisting of a finite number of elements. The AHP searches for the priorities representing the relative importance of the decision elements at each particular level. By suitable aggregation it finally calculates the priorities of the alternatives at the bottom level of the hierarchy. Their priorities are interpreted with respect to the overall goal at the top of the hierarchy, and elements at upper levels such as criteria, sub-criteria, etc. are used to mediate the comparison process. The elicitation process at the given level is performed by pairwise comparisons of all elements at given level of the hierarchy with respect to the elements of the upper level. If he/she prefers so, the DM may directly use a numerical value from the scale to express the ratio of elements' relative importance. By inserting numerical values into proper positions a PC matrix is created, and the role of the prioritization method is to extract the relative priorities—weights of all the compared alternatives, i.e., the rating of alternatives.

The values representing the preferences of the decision elements—alternatives can also be considered as the results of aggregation of the pairwise comparisons of a group of decision-makers and/or experts. Then the DM problem becomes the *group DM problem (GDM)*. The tournament ranking problem is another well-known application of pairwise comparisons (see e.g., [33]).

A crucial step in a DM process is the determination of a weighted ranking, i.e., a rating on a set $\mathscr{C} = \{c_1, c_2, ..., c_n\}$ of alternatives with respect to criteria or experts. A way to determine the rating is to start from a relation represented by the $n \times n$ PC matrix $A = \{a_{ij}\}$; each element of this matrix a_{ij} is a positive real number which expresses how much c_i is preferred to c_j.

The properties of the PC matrix depend on the various meanings given to the number a_{ij}, in particular, the $A = \{a_{ij}\}$ becomes: *multiplicative, additive, or fuzzy.* In the following sections we shall deal with these cases separately.

2.4 Multiplicative Pairwise Comparisons Matrices

Let us assume that the preferences over the set of alternatives $\mathscr{C} = \{c_1, c_2, \cdots, c_n\}$ is represented in the following way. Suppose that the intensities of the DM's preferences are given by an $n \times n$ matrix $A = \{a_{ij}\}$ with positive elements in such a way that, for each couple i and j, the entry of matrix A, a_{ij} indicates the ratio of preference intensity for alternative c_i to that of c_j. In other words, a_{ij} indicates that "c_i is a_{ij} times better than c_j." If, for example, c_i is 3 times better than c_j, then the goodness of c_j is $1/3$ with respect to the goodness of c_i. The elements of $A = \{a_{ij}\}$ satisfy the following reciprocity condition.

A positive $n \times n$ matrix $A = \{a_{ij}\}$ is *multiplicative-reciprocal (m-reciprocal)*, if

$$a_{ij}.a_{ji} = 1 \text{ for all } i, j \in \{1, \cdots, n\} , \tag{2.1}$$

or, equivalently:

$$a_{ji} = \frac{1}{a_{ij}} \text{ for all } i, j \in \{1, \cdots, n\} . \tag{2.2}$$

Reciprocity condition (2.1), or (2.2), seems to be natural in many decision situations. However, in some situations the assumption of reciprocity may be too restrictive. For example, in the case when elements under exchange are currencies, a coefficient a_{ij} is a rate of exchange of currency i in relation to currency j, then the presence of transaction costs makes the considered matrix $A = \{a_{ij}\}$ nonreciprocal, and, in particular, (2.1) no longer holds. However, in this book, we restrict our consideration only to the most popular reciprocal (particularly, m-reciprocal) preference matrices. The nonreciprocal case has been investigated, for example in [56]. Here, we shall assume that PC matrices are m-reciprocal.

It is very important to establish more important properties to be verified by PC matrices for designing good decision-making models. One of these properties is the so-called consistency property. Intuitively, the lack of consistency in decision-making can lead to incompatible conclusions; that is why it is important to study conditions under which consistency is satisfied (see e.g., [36, 44, 51] or [29]). On the other hand, perfect consistency is difficult to obtain in practice, especially when measuring preferences on a set with a large number of alternatives, e. g. $n > 10$. Clearly, the problem of consistency itself includes two special problems: when an expert, considered individually, is said to be consistent and, when the whole group of experts are considered consistent. In this section we will mainly focus on the first problem, assuming that experts preferences are expressed by means of a preference relation defined over a finite and fixed set of alternatives, expressed appropriately by a square matrix.

A positive $n \times n$ matrix $A = \{a_{ij}\}$ is *multiplicative-consistent* (or, *m-consistent*) [44, 93], if

$$a_{ik} = a_{ij}.a_{jk} \text{ for all } i, j, k \in \{1, \cdots, n\}, \tag{2.3}$$

or, equivalently

$$a_{ij}.a_{jk}.a_{ki} = 1 \text{ for all } i, j, k \in \{1, \cdots, n\} . \tag{2.4}$$

From (2.3), we obtain $a_{ii} = 1$ for all i, and also (2.3) implies (2.1), i.e., an m-consistent matrix is m-reciprocal (but not vice-versa).

Remark 2.1 Notice that $a_{ij} > 0$ for all elements of matrix A and m-consistency is not restricted to the Saaty's scale $S = \{1/9, 1/8, \cdots, 1/2, 1, 2, \cdots, 8, 9\}$. By the way, Saaty use this discrete scale only for evaluating the elements of A, while for other calculations with the matrix all positive numbers are allowed. Here, we extend this scale to the closed interval $S = [1/\sigma, \sigma]$, where $\sigma > 1$. If σ goes to infinity, we obtain the scale $S =]0, +\infty[$, i.e., the interval of all positive real numbers $\mathbf{R}_+$.

In practice, perfect consistency is difficult to obtain, particularly when evaluating preferences on a set with a large number of alternatives. If for some positive $n \times n$ matrix $A = \{a_{ij}\}$ and for some $i, j, k = 1, 2, \cdots, n$, the m-consistency condition (2.3) does not hold, then A is said to be *multiplicative-inconsistent* (or, *m-inconsistent*). In order to measure the grade of an inconsistency of a given matrix several instruments have been proposed in the literature (see [93]).

The following results give a characterization of m-consistent matrix (see also [93]).

Proposition 2.1 *Let $A = \{a_{ij}\}$ be a positive $n \times n$ matrix. A is m-consistent if and only if there exists a vector $w = (w_1, w_2, \cdots, w_n)$ with $w_i > 0$ for all $i \in \{1, \cdots, n\}$, and $\sum_{j=1}^{n} w_j = 1$ such that*

$$a_{ij} = \frac{w_i}{w_j} \text{ for all } i, j \in \{1, \cdots, n\} . \tag{2.5}$$

Proof (i) $A = \{a_{ij}\}$ be m-consistent. For $i \in \{1, \cdots, n\}$, set

$$v_i = (a_{i1}a_{i2}. \cdots .a_{in})^{\frac{1}{n}} . \tag{2.6}$$

Moreover, set

$$S = \sum_{i=1}^{n} v_i$$

and, finally, define

$$w_i = \frac{v_i}{S} \text{ for } i \in \{1, \cdots, n\} . \tag{2.7}$$

Then, for $i, j = 1, \cdots, n$, by m-reciprocity (2.1) and m-consistency (2.3) we obtain

$$\frac{w_i}{w_j} = (\frac{a_{i1}a_{i2}. \cdots .a_{in}}{a_{j1}a_{j2}. \cdots .a_{jn}})^{\frac{1}{n}} = ((a_{i1}a_{1j})(a_{i2}a_{2j}). \cdots .(a_{in}a_{nj}))^{\frac{1}{n}} = (a_{ij}.a_{ij}. \cdots .a_{ij})^{\frac{1}{n}} = a_{ij} .$$

Moreover, $\sum_{i=1}^{n} w_i = 1$, so consequently, (2.5) is true.

(ii) If (2.5) holds, then evidently (2.3) is satisfied, hence, $A = \{a_{ij}\}$ is m-consistent.   □

The following result follows from Remark 2.2.

Proposition 2.2 *Let $A = \{a_{ij}\}$ be a positive m-reciprocal $n \times n$ matrix. $A = \{a_{ij}\}$ is m-consistent if and only if there exists a vector $w = (w_1, w_2, \cdots, w_n)$ with $w_i > 0$ for all $i = 1, \cdots, n$, and*

$$\prod_{j=1}^{n} w_j = 1 , \tag{2.8}$$

such that

$$a_{ij} = \frac{w_i}{w_j} \text{ for all } i, j \in \{1, \cdots, n\} . \tag{2.9}$$

Proof (i) Let $A = \{a_{ij}\}$ be m-consistent, for $i \in \{1, \cdots, n\}$, set

$$w_i = (a_{i1}a_{i2}. \cdots .a_{in})^{\frac{1}{n}} . \tag{2.10}$$

Then, for $i, j \in \{1, \cdots, n\}$, by m-reciprocity (2.1) and m-consistency (2.3) we obtain

$$\frac{w_i}{w_j} = \left(\frac{a_{i1}a_{i2}. \cdots .a_{in}}{a_{j1}a_{j2}. \cdots .a_{jn}}\right)^{\frac{1}{n}} = ((a_{i1}a_{1j})(a_{i2}a_{2j}). \cdots .(a_{in}a_{nj}))^{\frac{1}{n}} = (a_{ij}.a_{ij}. \cdots .a_{ij})^{\frac{1}{n}} = a_{ij} .$$

Moreover,

$$\prod_{i=1}^{n} w_i = ((a_{11}a_{12}.\cdots.a_{1n})(a_{21}a_{22}.\cdots.a_{2n}).\cdots.(a_{n1}a_{n2}.\cdots.a_{n.n}))^{\frac{1}{n}} =$$

$$= ((a_{12}a_{21})(a_{13}a_{31}).\cdots.(a_{1n}a_{n1}).\cdots.(a_{2n}a_{n2}).\cdots.(a_{13}a_{31}).\cdots.(a_{n-1,n})(a_{n,n-1}))^{\frac{1}{n}} = 1 .$$

Consequently, (2.8) is true.

(ii) If (2.9) holds, then evidently (2.3) is satisfied. Hence, $A = \{a_{ij}\}$ is m-consistent. □

2.5 Methods for Deriving Priorities from Multiplicative Pairwise Comparison Matrices

Now, all matrices are $n \times n$-matrices with positive elements and all vectors are n-dimensional vectors with positive elements.

Let $\mathscr{A}^{\times}$ be the set of all m-reciprocal matrices with positive elements. Then $\mathscr{A}^{\times}$ is a multiplicative group under componentwise multiplication. If $A = \{a_{ij}\} \in \mathscr{A}^{\times}$ and $B = \{b_{ij}\} \in \mathscr{A}^{\times}$, then $C = A \times B = \{a_{ij}.b_{ij}\} \in \mathscr{A}^{\times}$. Here, by $\times$ we denote the group operation, i.e., the componentwise multiplication of matrices (in contrast to the usual matrix multiplication). Similarly, let $\mathscr{A}^{\times c}$ be the set of all m-consistent matrices and $w^{\times}$ be the set of all vectors $w = (w_1, w_2, \cdots, w_n)$ with positive elements such that

$$\prod_{j=1}^{n} w_j = 1.$$

Both $\mathscr{A}^{\times c}$ and $w^{\times}$ are multiplicative groups under componentwise multiplication. Moreover, $\mathscr{A}^{\times c}$ is a subgroup of $\mathscr{A}^{\times}$ and by Proposition 2.2, $\mathscr{A}^{\times c}$ is isomorphic to $w^{\times}$.

Let $\mathscr{F}^{\times}$ be the set of all mappings from $\mathscr{A}^{\times}$ into $w^{\times}$, $f \in \mathscr{F}^{\times}$.

Vector $w = (w_1, w_2, \cdots, w_n) \in w^{\times}$ is called the *m-priority vector associated with* $A = \{a_{ij}\}$, or, the *m-priority vector of* $A = \{a_{ij}\}$, if there exists $f \in \mathscr{F}^{\times}$ such that $w = f(A)$ and

$$a_{ij} = w_i/w_j \text{ for all } i, j \in \{1, \cdots, n\} . \tag{2.11}$$

Then w is called the *m-consistency vector*.

Here, the mapping f defines how the m-priority vector "is calculated" from the elements a_{ij} of A. If f is defined by (2.10), then (2.11) is satisfied if and only if A is m-consistent.

In the DM problem given by $\mathscr{C} = \{c_1, c_2, \cdots, c_n\}$ and $A = \{a_{ij}\}$, the rating of the alternatives in $\mathscr{C}$ is determined by the priority vector $w = (w_1, w_2, \cdots, w_n)$ of A. This vector, if normalized, is called the *vector of weights*. Hence, each element

w_i of the m-priority vector w is interpreted as the *relative importance of alternative* c_i. The ranking of alternatives is defined as follows:

$$c_i \succ c_j \text{ if } w_i > w_j \text{ for all } i, j \in \{1, \cdots, n\} .$$

Therefore, the alternatives $c_1, c_2, \cdots, c_n$ in X can be ranked/rated by their relative importance.

Generally, the PC matrix $A = \{a_{ij}\}$ is not m-consistent, so the priority vector cannot be an m-consistency vector. We face the problem of how to measure the inconsistency of the PC matrix. In order to solve this problem, we shall define special consistency indexes based on the corresponding priority vectors.

In the next section, six popular methods for deriving priority vector are discussed. Our choice of methods is based on [32], where 18 different methods for deriving priority vectors from PC matrices under a common framework of effectiveness have been discussed. Later on, in [68], C.-C. Lin published a revised framework for deriving preference values from pairwise comparison matrices together with some new simulation. In [99] B. Srdjevic combined different prioritization methods in AHP. Here, the concept of minimizing the aggregated deviation and so-called *correctness in error-free cases* are presented. Some comparison calculations demonstrate the corresponding results of the presented methods. For each method we also derive an associated consistency index—a tool for measuring the grade of inconsistency of the given PC matrix. However, in Sect. 2.7 we propose a new method of deriving priority vectors satisfying some desirable properties by solving a special optimization problem.

2.5.1 Eigenvector Method (EVM)

Historically, one of the oldest methods for deriving priorities from the multiplicative preference matrix is the eigenvector method (EVM) (see [93]). This method has an intuitive background but it is now based on Perron–Frobenius theory which is known in several versions (see e.g., [45]), where you can find the proof, which is nontrivial. The Perron theorem stated below for positive matrices describes some of the remarkable properties enjoyed by the eigenvalues and eigenvectors. Later, Frobenius extended this theorem also for irreducible nonnegative matrices (see e.g., [45]).

Theorem 2.1 (Perron) *Let $A = \{a_{ij}\}$ be a positive $n \times n$ matrix. Then the spectral radius of A, $\rho(A)$, is a real eigenvalue, which has a positive (real) eigenvector $w = (w_1, w_2, \cdots, w_n)$. This eigenvalue called the* principal eigenvalue *of A is simple (it is not a multiple root of the characteristic equation), and its eigenvector, called the* principal eigenvector *of A, is unique up to a multiplicative constant.*

By EVM, the priority vector of a positive $n \times n$ matrix $A = \{a_{ij}\}$ is defined as the normalized principal eigenvector $w = f(A)$ given by Perron theorem. Hence, the rating of the alternatives in $\mathscr{C}$ is determined by the priority vector $w = (w_1, w_2, \cdots, w_n)$, with $w_i > 0$, for all $i \in \{1, \cdots, n\}$, such that $\sum_{i=1}^{n} w_i = 1$, satisfying $Aw = \rho(A)w$. Since the element of the priority vector w_i is interpreted as the relative importance of alternative c_i, the alternatives $c_1, c_2, \cdots, c_n$ in $\mathscr{C}$ are rated by their relative importance.

According to Saaty [93], an inconsistency grade of a positive m-reciprocal $n \times n$ matrix $A = \{a_{ij}\}$ can be measured by the m-EV-consistency index $CI_{mEV}(A)$ defined as

$$CI_{mEV}(A) = \frac{\rho(A) - n}{n - 1}, \qquad (2.12)$$

where $\rho(A)$ is the spectral radius of A (i.e., the principal eigenvalue of A). The proof of the following result can be found in [93].

Proposition 2.3 *If $A = \{a_{ij}\}$ is an $n \times n$ positive m-reciprocal matrix, then $CI_{mEV} \geq 0$. Moreover, A is m-consistent if and only if $CI_{mEV}(A) = 0$.*

To provide an inconsistency measure independently of the dimension n of the matrix A, Saaty in [93] proposed the consistency ratio CR_{mEV}. In order to distinguish it here from the other inconsistency measures, we shall call it the m-EV-consistency ratio. This is obtained by taking the consistency index CI_{mEV} to its mean value R_{mEV}, i.e., the mean value of $CI_{mEV}(A)$ of positive m-reciprocal matrices of dimension n, whose entries are uniformly distributed random variables on the interval $[1/9; 9]$, i.e.,

$$CR_{mEV} = \frac{CI_{mEV}}{R_{mEV}}. \qquad (2.13)$$

The following table relates the dimension n of the positive m-reciprocal matrix in the first row to its corresponding mean value $R_{mEV}(n)$, $n = 3, 4, \ldots, 10$ in the second row.

n	3	4	5	6	7	8	9	10
$R_{mEV}(n)$	0.58	0.90	1.12	1.24	1.32	1.41	1.45	1.49

Now, we define an alternative m-consistency index based on the distance of a positive m-reciprocal $n \times n$ matrix $A = \{a_{ij}\}$ to the *ratio matrix* $W = \{\frac{w_i}{w_j}\}$. We define the m-EV-consistency index $I_{mEV}(A)$ as

$$I_{mEV}(A) = \sum_{1 \leq i < j \leq n} \left(a_{ij} - \frac{w_i^*}{w_j^*} \right)^2, \qquad (2.14)$$

where $w^* = (w_1^*, w_2^*, \cdots, w_n^*)$ is the principal eigenvector from Peron's Theorem 2.1 above. We obtain the result parallel to Proposition 2.3 as follows.

Proposition 2.4 *If $A = \{a_{ij}\}$ is an $n \times n$ positive m-reciprocal matrix, then $I_{mEV} \geq 0$. Moreover, A is m-consistent if and only if $I_{mEV}(A) = 0$.*

For the m-EV-consistency ratio it was proposed an estimation of a 10% threshold of CR_{mEV}. In other words, a pairwise comparison matrix could be acceptable (in a DM process) if its m-consistency ratio does not exceed 0.1 (see [93]).

The following theorem, which follows directly from Perron's theorem, is useful for practical computations of the priority vector and also for calculating the m-consistency index (2.12), and/or m-consistency ratio (2.13). The proof of the theorem, named by H. Wielandt, can be found, for example, in [93].

Theorem 2.2 (Wielandt) *Let $A = \{a_{ij}\}$ be a positive $n \times n$ matrix, $e = (1, 1, \cdots, 1)$ be an n-vector. Then the principal eigenvector $w = (w_1, w_2, \cdots, w_n)$ corresponding to the principal eigenvalue of A is as follows:*

$$w = \lim_{k \to \infty} \frac{A^k e}{e^T A^k e} . \tag{2.15}$$

Notice that $A^k e$ is the vector of the row sums of the k-powered matrix A^k, whereas $e^T A^k e$ is the sum of all elements of A^k. Applying Theorem 2.2 we can easily calculate the priority vector of the matrix A (consequently, the principal eigenvalue and the consistency ratio) by formula (2.15), simply by calculating the $2k$-th powers of A, i.e., A, A^2, $(A^2)^2$, $\cdots$. Here, we apply the usual matrix multiplication, not a componentwise one, as before. Such calculations can be performed easily using Excel.

It was shown by various researchers (see e.g., [7]), that for small deviations of a_{ij} around the consistent ratios w_i/w_j, i.e., for small deviations $a_{ij} - w_i/w_j$, EVM gives a reasonably good approximation of the priority vector. However, when the deviations are not very small, it is generally accepted that the corresponding priority vector is not satisfactory. This is the main reason why we look for other methods for deriving priority vectors.

2.5.2 Arithmetic Mean Method (AMM)

The arithmetic mean method (AMM) is a heuristic method which is based on Theorem 2.2. To obtain the priority vector $w = (w_1, w_2, \cdots, w_n)$ by this method it is enough to divide the elements of each column of matrix $A = \{a_{ij}\}$ by the sum of that column (i.e., to normalize the column), then add the elements in each resulting row and finally divide this sum by n, the number of elements in the row. The element of the priority vector is given by

$$w_i = \frac{1}{n} \sum_{j=1}^{n} \frac{a_{ij}}{\sum_{k=1}^{n} a_{kj}}, \text{ for all } i = 1, \cdots, n . \tag{2.16}$$

An m-inconsistency measure of a positive m-reciprocal $n \times n$ matrix $A = \{a_{ij}\}$ with respect to AMM is given by the *m-AM-consistency index* $I_{mAM}(A)$ defined as

$$I_{mAM}(A) = \sum_{1 \leq i < j \leq n} \left(a_{ij} - \frac{w_i^*}{w_j^*} \right)^2 , \qquad (2.17)$$

where $w^* = (w_1^*, w_2^*, \cdots, w_n^*)$ is given by (2.16). We obtain the result parallel to Proposition 2.3 as follows.

Proposition 2.5 *If $A = \{a_{ij}\}$ is an $n \times n$ positive m-reciprocal matrix, then $I_{mAM} \geq 0$ and A is m-consistent if and only if $I_{mAM}(A) = 0$. Moreover, if A is m-consistent, then the AMM gives the same priority vector as EVM.*

Example 2.1 Let $\mathscr{C} = \{c_1, c_2, c_3, c_4\}$ be the set of alternatives and assume that the intensities of the DM's preferences are given by an 4×4 matrix $A = \{a_{ij}\}$:

$$A = \begin{pmatrix} 1 & \frac{1}{2} & 3 & 4 \\ 2 & 1 & \frac{1}{5} & 6 \\ \frac{1}{3} & 5 & 1 & \frac{1}{7} \\ \frac{1}{4} & \frac{1}{6} & 7 & 1 \end{pmatrix} .$$

We calculate the priority vector w^* by AMM, (2.16), as follows:

$$w^* = (w_1^*, w_2^*, w_3^*, w_4^*) = (0.25, 0.31, 0.24, 0.20) .$$

According to the priority vector w^* we obtain the same ranking: $c_2 \succ c_1 \succ c_3 \succ c_4$ and the m-AM-consistency index $I_{mAM}(A) = 33.74$.

Now, we compute the priority vector w^{**} by EV method as follows:

$$w^{**} = (0.25, 0.29, 0.22, 0.24) ,$$

m-AM-consistency index $I_{mEV}(A) = 67.35$. According to the priority vector w^{**} we obtain the ranking: $c_2 \succ c_1 \succ c_4 \succ c_3$.

We can see that the AMM and EVM give different priority vectors and different ranking of alternatives. Moreover, $I_{mEV}(A) = 67.35 > 33.74 = I_{mAM}(A)$.

In [93] Saaty suggested AMM as a simplified and tractable version of EVM. In [68], this method is ranked among the three best ones in simulation experiments, together with the EVM and LLSM, see below. Popularity and wide use in practice AMM are due to its extreme simplicity. Although considered inferior, it significantly outperforms more sophisticated methods, as is demonstrated in [68, 99].

2.5.3 Least Squares Method (LSM)

The least squares (LS) method minimizes the L_2 distance function defined for elements of the unknown priority vector $w = (w_1, w_2, \cdots, w_n)$ and known elements a_{ij} of the matrix A by solving the following constrained nonlinear optimization problem:

$$\sum_{i,j=1}^{n} \left(a_{ij} - \frac{w_i}{w_j} \right)^2 \longrightarrow \min; \tag{2.18}$$

subject to

$$\sum_{j=1}^{n} w_j = 1, w_i \geq \varepsilon > 0 \text{ for all } i \in \{1, \cdots, n\},$$

(where ε is a preselected sufficiently small positive number).

As the above nonlinear optimization problem may have multiple solutions it is advantageous to convert this problem to the following one:

$$\sum_{1 \leq i < j \leq n} \left(a_{ij} w_j - w_i \right)^2 \longrightarrow \min; \tag{2.19}$$

subject to

$$\sum_{j=1}^{n} w_j = 1, w_i \geq \varepsilon > 0 \text{ for all } i \in \{1, \cdots, n\}. \tag{2.20}$$

Optimization problem (2.19), (2.20) which is referred to as *weighted LSM* is transformed into a system of linear equations by differentiating the Lagrangian of (2.19) and equalizing it to zero. It is shown in [10] that in this way the WLSM provides a unique and strictly positive solution $w^* = (w_1^*, w_2^*, \cdots, w_n^*)$. This solution is set as the priority vector of A. If A is m-consistent then it is clear that $A = \{w_i^*/w_j^*\}$, hence, w^* is a consistent vector.

An m-inconsistency measure of a positive m-reciprocal $n \times n$ matrix $A = \{a_{ij}\}$ with respect to LSM is given by the *m-LS-consistency index* $I_{mLS}(A)$ defined as

$$I_{mLS}(A) = \sum_{1 \leq i < j \leq n} \left(a_{ij} - \frac{w_i^*}{w_j^*} \right)^2, \tag{2.21}$$

where $w^* = (w_1^*, w_2^*, \cdots, w_n^*)$ is the optimal solution of (2.19), (2.20). The proof of the following proposition which is parallel to Proposition 2.3 is straightforward.

Proposition 2.6 *If $A = \{a_{ij}\}$ is a positive m-reciprocal matrix, then $I_{mLS}(A) \geq 0$. Moreover, A is m-consistent if and only if $I_{mLS}(A) = 0$.*

Example 2.2 Let $\mathscr{C} = \{c_1, c_2, c_3, c_4\}$ be the set of alternatives and assume that the intensities of the DM's preferences are given by an 4×4 matrix $A = \{a_{ij}\}$ from Example 2.1, i.e.,

$$A = \begin{pmatrix} 1 & \frac{1}{2} & 3 & 4 \\ 2 & 1 & \frac{1}{5} & 6 \\ \frac{1}{3} & 5 & 1 & \frac{1}{7} \\ \frac{1}{4} & \frac{1}{6} & 7 & 1 \end{pmatrix} .$$

We compute the priority vector w^+ by the LSQM, (2.19), as follows:

$$w^+ = (0.38, 0.40, 0.14, 0.08) .$$

According to the priority vector w^+ we obtain the ranking: $c_2 \succ c_1 \succ c_3 \succ c_4$, the same ranking as by AMM. Moreover, $I_{mLS}(A) = 15.12$.

2.5.4 Logarithmic Least Squares Method (LLSM)/Geometric Mean Method (GMM)

The logarithmic least squares method (LLSM), presented in [7, 34], makes use of the multiplicative properties of the PC matrix $A = \{a_{ij}\}$. LLSM assumes the minimization of the sum of the logarithmic squared deviations from given elements a_{ij} of A, i.e.,

$$\sum_{1 \leq i < j \leq n} \left(\ln a_{ij} - \ln \frac{w_i}{w_j} \right)^2 \longrightarrow \min; \tag{2.22}$$

subject to

$$\prod_{j=1}^{n} w_j = 1, w_i \geq 0 \text{ for all } i = 1, \cdots, n . \tag{2.23}$$

As can be easily demonstrated (see e.g., [34]), the optimal solution of problem (2.22), (2.23) is always unique and can be found simply as the geometric mean of the rows of the reciprocal PC matrix, provided that the set of given pairwise comparisons is complete, i.e., the number of judgments is $n(n-1)/2$. The elements of the priority vector—the optimal solution of problem (2.22), (2.23) $w^* = (w_1^*, w_2^*, \cdots, w_n^*)$—are defined as the geometric mean of the row elements of A:

$$w_i^* = \left(\prod_{j=1}^{n} a_{ij} \right)^{1/n} \text{ for all } i \in \{1, \cdots, n\} . \tag{2.24}$$

That is why LLSM is also referred to as the *geometric mean method* (GMM).

Notice that constraint (2.23) can be substituted by the usual normalization constraint as follows:

$$\sum_{j=1}^{n} w_j = 1, w_i \geq 0 \text{ for all } i \in \{1, \cdots, n\} . \tag{2.25}$$

Clearly, if the vector w^* is an optimal solution of problem (2.22), (2.23), then $v^* = c.w^*$ is an optimal solution of problem (2.22), (2.25), for a suitable $c > 0$. The opposite assertion is also true. The vector v^* is an optimal solution of problem (2.22), (2.25) if $w^* = d.v^*$ is an optimal solution of problem (2.22), (2.23) for a suitable $d > 0$.

Let $A = \{a_{ij}\}$ be a PCM, $w^* = (w_1^*, w_2^*, \cdots, w_n^*)$ be the optimal solution of (2.22) given by (2.24), and set

$$e_{ij}^m = a_{ij} \frac{w_j^*}{w_i^*} \text{ for all } i \in \{1, \cdots, n\} .$$

We denote

$$E_A^m = \{e_{ij}^m\} = \{a_{ij} \frac{w_j^*}{w_i^*}\}, W_A^m = \{\frac{w_i^*}{w_j^*}\} .$$

Here, $E_A^m = \{e_{ij}^m\}$ is called the *m-error matrix of* $A = \{a_{ij}\}$. By (2.24) we obtain

$$e_{ij}^m = \left(\prod_{k=1}^{n} a_{ij} a_{jk} a_{ki}\right)^{1/n} .$$

Notice that $W_A^m = \{\frac{w_i^*}{w_j^*}\} \in \mathscr{A}^{\times c}$, i.e., W_A^m is m-consistent. Let us denote $(W_A^m)^{-1} = \{\frac{w_j^*}{w_i^*}\}$, then evidently

$$A = E_A^m \times (W_A^m)^{-1} . \tag{2.26}$$

An m-consistency measure of a positive m-reciprocal $n \times n$ matrix $A = \{a_{ij}\}$ with respect to LLSM (GMM) is given by the *m-consistency index* $I_{mGM}(A)$ defined as the minimal value of the objective function (2.22) as

$$I_{mGM}(A) = \sum_{1 \leq i < j \leq n} \ln^2(e_{ij}) , \tag{2.27}$$

The proof of the following proposition which is parallel to Proposition 2.6 is straightforward.

Proposition 2.7 *If* $A = \{a_{ij}\}$ *is a positive m-reciprocal matrix, then* $I_{mGM}(A) \geq 0$. *Moreover,* A *is m-consistent if and only if* $I_{mGM}(A) = 0$.

Proposition 2.7 enables us to distinguish between m-consistent and m-inconsistent matrices but it is insufficient to determine the degree of m-consistency of m-inconsistent matrices. In fact, a statement of the type: *A is less m-consistent than B if* $I_{mGM}(A) > I_{mGM}(B)$ is not meaningful when A and B are of different dimensions. Moreover, a cut-off rule of the type: *A is close enough to being m-consistent if* $I_{mGM}(A) \leq \alpha$ for some fixed positive constant α, independent of the dimension n, does not appear to be meaningful.

A natural way to construct a better measure than the above defined m-consistency index I_{mGM} that would preserve the abovementioned properties and addresses its deficiencies is to consider the *relative m-error* $RE_m(A)$ *of* $A = \{a_{ij}\}$ defined as the normalized I_{mGM}, see [7]

$$RE_m(A) = \frac{\sum_{1 \leq i < j \leq n} \ln^2(e_{ij})}{\sum_{1 \leq i < j \leq n} \ln^2(a_{ij})}, \quad (2.28)$$

Proposition 2.8 *If* $A = \{a_{ij}\}$ *is a positive m-reciprocal matrix, then*

$$0 \leq RE_m(A) \leq 1. \quad (2.29)$$

Proof As the error component minimizes the sum of squares (2.22) we obtain

$$\sum_{1 \leq i < j \leq n} \ln^2(a_{ij} \frac{w_j^*}{w_i^*}) \leq \sum_{1 \leq i < j \leq n} \ln^2(a_{ij} \frac{1}{1}) = \sum_{1 \leq i < j \leq n} \ln^2(a_{ij}),$$

hence, (2.29) is satisfied. □

By Proposition 2.8, the values of the relative error belong to the unit interval $[0, 1]$ regardless of n, the dimension of A. Using this measure, we may compare the consistency of matrices of different dimensions and justify the use of cut-off rules, accepting A as sufficiently consistent if, for example, its relative error satisfies $RE_m(A) \leq 0.1$.

A matrix A is called *totally m-inconsistent* if its relative m-error is maximal, i.e., $RE_m(A) = 1$.

The relative error measures the relative grade of inconsistency of A. Now, we define a new index which will measure the relative grade of consistency of a matrix A by the matrix W_A.

The *relative m-consistency index* $RC_m(A)$ *of* $A = \{a_{ij}\}$ is defined as follows (see [7]):

$$RC_m(A) = \frac{\sum_{1 \leq i < j \leq n} \ln^2(\frac{w_i^*}{w_j^*})}{\sum_{1 \leq i < j \leq n} \ln^2(a_{ij})}, \quad (2.30)$$

where $w^* = (w_1^*, w_2^*, \cdots, w_n^*)$ is given by (2.24), the optimal solution of (2.22), (2.23).

Proposition 2.9 *If $A = \{a_{ij}\}$ is a positive m-reciprocal matrix, then*

$$RC_m(A) + RE_m(A) = 1 \ . \tag{2.31}$$

The proof of this proposition will be given later in Sect. 2.5 in connection with additive preference matrices and the logarithmic/exponential isomorphism between multiplicative and additive PC matrices.

Proposition 2.10 *$A = \{a_{ij}\} \in \mathscr{A}^\times$ is totally m-inconsistent if and only if the row products of A are all ones, i.e.,*

$$\prod_{j=1}^{n} a_{ij} = 1 \ for \ all \ i \in \{1, \cdots, n\} \ . \tag{2.32}$$

Proof (i) Let $A = \{a_{ij}\}$ be totally m-inconsistent, i.e., $RE_m(A) = 1$, hence, by Proposition 2.9 we obtain

$$RC_m(A) = \frac{\sum_{1 \le i < j \le n} \ln^2(\frac{w_i^*}{w_j^*})}{\sum_{1 \le i < j \le n} \ln^2(a_{ij})} = 0 \ .$$

Therefore, $w_i^* = w_j^*$ for all $i, j = 1, \cdots, n$ and

$$\sum_{1 \le i < j \le n} \ln^2(\frac{w_i^*}{w_j^*}) = 0 \ .$$

Consequently, (2.32) holds.

(ii) Let (2.32) be satisfied. Then by (2.24) $w_i^* = w_j^* = 1$ for all $i, j \in \{1, \cdots, n\}$ and then

$$\sum_{1 \le i < j \le n} \ln^2(a_{ij} \cdot \frac{w_j^*}{w_i^*}) = \sum_{1 \le i < j \le n} \ln^2(a_{ij}) \ .$$

Hence, $RE_m(A) = 1$.   □

Example 2.3 Let $\mathscr{C} = \{c_1, c_2, c_3\}$ be the set of 3 alternatives and assume that the intensities of preferences are given by an 3×3 matrix $A = \{a_{ij}\} \in \mathscr{A}^\times$ defined as

$$A = \begin{pmatrix} 1 & 2 & \frac{1}{16} \\ \frac{1}{2} & 1 & 8 \\ 16 & \frac{1}{8} & 1 \end{pmatrix} \ .$$

We compute $w^* = (w_1^*, w_2^*, w_3^*)$ as the row geometric averages of A, then we compute $B = \{\ln^2(\frac{w_i^*}{w_j^*})\}$ and $C = \{\ln^2(a_{ij})\}$ as follows:

$$w^* = (w_1^*, w_2^*, w_3^*) = ((1 \cdot 2 \cdot \frac{1}{16})^{\frac{1}{3}}, (\frac{1}{2} \cdot 1 \cdot 8)^{\frac{1}{3}}, (16 \cdot \frac{1}{8} \cdot 1)^{\frac{1}{3}}) = (0.50, 1.59, 1.26) ,$$

the normalized priority vector $v^* = \frac{1}{3.35} \cdot w^* = (0.15, 0.47, 0.38)$,

$$B = \{b_{ij}\} = \begin{pmatrix} 0 & 1.33 & 0.85 \\ 0.48 & 0 & 0.05 \\ 0.85 & 0.05 & 0 \end{pmatrix} ,$$

$$C = \{c_{ij}\} = \begin{pmatrix} 0 & 0.48 & 7.69 \\ 0.48 & 0 & 4.32 \\ 7.69 & 4.32 & 0 \end{pmatrix} .$$

Then we calculate

$$RC_m(A) = \frac{\sum_{1 \leq i < j \leq 3} b_{ij}}{\sum_{1 \leq i < j \leq 3} c_{ij}} = \frac{2.24}{12.49} = 0.18 .$$

Moreover, by (2.31) we have $RE_m(A) = 1 - RC_m(A) = 1 - 0.18 = 0.82$. We conclude that the relative m-consistency of A is 18% and the relative m-error of A is 82%. Hence, the m-inconsistency of A is relatively high.

According to the priority vector v^* we obtain the ranking: $c_2 \succ c_3 \succ c_1$.

2.5.5 Fuzzy Programming Method

The fuzzy programming method (FPM) proposed by Mikhailov in [75] firstly states that if reciprocal matrix $A = \{a_{ij}\}$ is m-consistent, then the system of $m = n(n - 1)/2$ linear equations:

$$a_{ij}w_j - w_i = 0 \text{ for all } i, j \in \{1, \cdots, n\}, i < j ,$$

can be represented as follows:

$$C_j w = 0, \ j \in \{1, \cdots, m\} , \tag{2.33}$$

where by C_j we denote the jth row vector of coefficients of the matrix of the system (2.33).

If A is inconsistent, it is desirable to find all values of vector w, such that (2.33) is approximately satisfied, i.e., $Cw \approx 0$.

The FPM represents (2.33) geometrically as an intersection of fuzzy hyperlines and transforms the prioritization problem to an optimization one, determining the values of the priorities that correspond to the point with the highest *measure of intersection*. In this way the prioritization problem is reduced to a fuzzy programming problem that can be solved as a standard linear program:

$$\mu \longrightarrow \max;$$

subject to

$$\mu.d_j^+ + C_j.w \le d_j^+, \ j \in \{1, \cdots, m\} \ ,$$

$$\mu.d_j^- - C_j.w \le d_j^-, \ j \in \{1, \cdots, m\} \ ,$$

$$\sum_{i=1}^n w_i = 1, w_i \ge 0 \text{ for all } i \in \{1, \cdots, n\} \ .$$

Here, the positive values of the left-and right-tolerance parameters d_j^+ and d_j^- represent the admissible interval of approximate satisfaction of the crisp equality (2.33).

Other methods for deriving priorities from multiplicative PC matrices can be found in the literature (see e.g., [32, 99] or [68]).

2.6 Desirable Properties of the Priority Vector

As will be shown in the following text of this chapter, some inconsistent pairwise comparisons matrices violate the fundamental selection (FS) condition of multiple criteria decision-making: the "best" alternative is selected from the set of non-dominated alternatives, while this set is nonempty. Inconsistent PCMs that violate this natural condition should be viewed as logically flawed and should not be used for the derivation of weights of alternatives or other objects. Other PCMs may violate the preservation of the order of preferences (the so-called POP condition), or preservation of the intensity of preference (the so-called POIP condition), see Bana e Costa and Vansnick [5]. PCMs in these two categories are in fact very frequent [73], so they might be considered acceptable for derivation of weights of alternatives or criteria in practice. However, a decision-maker should be aware of their logical limitations.

Furthermore, a new nonlinear optimization problem is proposed for generating a priority vector (weights of alternatives, criteria, or other alternatives). The method is designed to find a priority vector so that all three of the aforementioned properties are satisfied, hence providing a more logical solution than the classical eigenvalue (EV), or the geometric mean (GM) methods.

Definition 2.1 Let $A = \{a_{ij}\}$ be the PC matrix based on the set of alternatives $\mathscr{C} = \{c_1, c_2, \cdots, c_n\}$. We say that an alternative c_i *dominates alternative* c_j or equivalently, that an alternative c_j *is dominated by alternative* c_i, if

$$a_{ij} > 1. \tag{2.34}$$

If a given alternative is not dominated by any other alternative, then such alternative is called the *non-dominated alternative*. The set of all non-dominated alternatives in $\mathscr{C}$ with respect to matrix A is denoted by $ND(A)$.

By (2.34) we obtain

$$ND(A) = \{c_j \in \mathscr{C}| \text{ and there is no } i \in \{1, ..., n\} : a_{ij} > 1\}. \tag{2.35}$$

The following proposition gives a sufficient condition for the existence of a non-dominated alternative in $\mathscr{C}$. This property is well-known in the theory of graphs (see e.g., [12]).

Proposition 2.11 *Let $A = \{a_{ij}\}$ be the PC matrix based on the set of alternatives $\mathscr{C} = \{c_1, c_2, \cdots, c_n\}$. If there is no cycle of pairs of indexes*

$$(k_1, k_2), (k_2, k_3), \cdots, (k_{n-1}, k_n), (k_n, k_1),$$

where $k_i \in \{1, \cdots, n\}$, such that $\{k_1, k_2, \cdots, k_n\}$ is a permutation of $\{1, \cdots, n\}$ with

$$a_{k_i, k_{i+1}} > 1, i \in \{1, \cdots, n - 1\}, \text{ and } a_{k_n, k_1} > 1, \tag{2.36}$$

then $ND(A)$ is nonempty.

Definition 2.2 Let $A = \{a_{ij}\}$ be the PC matrix based on the set of alternatives $\mathscr{C} = \{c_1, c_2, \cdots, c_n\}$. Assume that $ND(A)$ is nonempty.

Let $w = (w(c_1), ..., w(c_n))$ be the priority vector (i.e., vector of weights) associated to A. We say that the *fundamental selection condition (FSC) is satisfied with respect to A and w*, if the maximal weight of the priority vector is associated with a non-dominated alternative.

Equivalently, we say that w satisfies the FSC with respect to A, if for some $i^* \in \{1, \cdots, n\}$

$$c_{i^*} \in ND(A) \text{ and } w(c_{i^*}) = \max\{w(c_j)| j \in \{1, \cdots, n\}\}. \tag{2.37}$$

Alternatively, we say that A satisfies FSC with respect to w.

Other suitable concepts of PCMs called conditions of order preservation (COP) were introduced and investigated, see [5], particularly preservation of order preference condition (POP) and preservation of order of intensity of preference condition (POIP) (see e.g., [61]).

Definition 2.3 Let $A = \{a_{ij}\}$ be a pairwise comparison matrix, and let $w = (w_1, \cdots, w_n)$ be a priority vector associated to A. A PC matrix A is said to satisfy *preservation of order preference condition (POP condition) with respect to priority vector w* if

$$a_{ij} > 1 \Rightarrow w_i > w_j. \tag{2.38}$$

The following example demonstrates a "sufficiently" consistent PCM where Saaty's consistency index is less than 0.1, but the FS and POP conditions are, however, not met.

Example 2.4 Consider the set of four alternatives $\mathscr{C} = \{c_1, c_2, c_3, c_4\}$, and the corresponding PC matrix A given as follows:

$$A = \begin{pmatrix} 1 & \frac{3}{2} & 2 & 2 \\ \frac{2}{3} & 1 & 4 & 4 \\ \frac{1}{2} & \frac{1}{4} & 1 & 1 \\ \frac{1}{2} & \frac{1}{4} & 1 & 1 \end{pmatrix},$$

From the first row of PC matrix A, alternative c_1 clearly dominates the other three alternatives. Hence, c_1 is non-dominated. Saaty's consistency index CI and consistency ratio CR are: $CI = 0.052$, $RI = 0.089$, $CR = CI/RI = 0,058$. According to Saaty, see [93], for a 4×4 PCM, inconsistency is acceptable if $CR < 0.08$, (for an $n \times n$ PCM, $n > 4 : CR < 0.1$). Hence, inconsistency of A is acceptable and the priority vector (additively normalized) is generated by EVM as follows:

$$w_{EV} = (0.350, 0.396, 0.127, 0.127).$$

The weights of all alternatives (the priority vector w additively normalized) derived by GMM are as follows:

$$w_{GM} = (0.344, 0.396, 0.130, 0.130).$$

According to both EVM and GMM, the alternative with the highest weight is alternative c_2. This alternative is, however, dominated by alternative c_1, which is the only non-dominated one. Therefore, both $\mathbf{w_{EV}}$ and also $\mathbf{w_{GM}}$ violate FSC with respect to A, even though the consistency ratio $CR = 0.058$ is below Saaty's threshold of 0.08.

Moreover, the POP condition is violated, too, as $a_{12} = \frac{3}{2} > 1$ and $\frac{w_1}{w_2} = \frac{0.344}{0.396} = 0.866 < 1$.

The following proposition says that the POP condition is stronger than the FS condition, i.e., POP condition implies FSC. The opposite assertion does not hold as it is demonstrated in Example 2.4.

Proposition 2.12 *Let $A = \{a_{ij}\}$ be a pairwise comparison matrix, and let $w = (w_1, \cdots, w_n)$ be a priority vector associated to A. If A satisfies the POP condition with respect to w, then A satisfies the FS condition with respect to w.*

Proof Assume that c_{i_0} is non-dominated, and w_{j_0} is the maximal weight, $i_0 \neq j_0$. If c_{j_0} is non-dominated, then the proposition holds. On the other hand, if c_{j_0} is dominated, then there is a c_{k_0} which dominates c_{j_0}, so $a_{k_0 j_0} > 1$. By the POP condition we have $w_{k_0} > w_{j_0}$, a contradiction with the assumption that w_{j_0} is a maximal weight.

Hence, the FS condition is satisfied. □

In the following definition we introduce the POIP condition [5, 61]. Usually, the DMs ask that the POIP condition is satisfied if the POP condition is also satisfied, however, both conditions are independent of one another as it is demonstrated below by examples. It will be shown that one condition is satisfied while, simultaneously, the other one is violated.

Definition 2.4 Let $A = \{a_{ij}\}$ be a pairwise comparison matrix, and let $w = (w_1, \cdots, w_n)$ be a priority vector associated with A. A PC matrix A is said to satisfy the *preservation of order intensity preference condition (POIP condition) with respect to vector w* if

$$a_{ij} > 1, a_{kl} > 1, \text{ and } a_{ij} > a_{kl} \Rightarrow \frac{w_i}{w_j} > \frac{w_k}{w_l}. \quad (2.39)$$

Definition 2.5 A PC matrix A is said to satisfy the *reliable preference (RP) condition with respect to priority vector w* if

$$a_{ij} > e \Rightarrow w_i > w_j, \quad (2.40)$$

$$a_{ij} = e \Rightarrow w_i = w_j. \quad (2.41)$$

Remark 2.2 From Definition 2.5 it is clear that the RP condition is stronger than the POP condition. In other words, if the RP condition is satisfied, then the POP condition is satisfied, too.

Remark 2.3 Let $A = \{a_{ij}\}$ be a consistent pairwise comparison matrix, and let $w = (w_1, \cdots, w_n)$ be a priority vector associated with A satisfying (2.11). Then it is obvious that FS, POP, and POIP conditions are satisfied. Moreover, for a consistent pairwise comparison matrix, it is well known, see e.g., [92], that the priority vector satisfying (2.11) can be generated by either EVM or by GMM.

A PC matrix A from Example 2.4 violates the POP condition with respect to priority vector w generated by the GM method (2.24). The following proposition gives a sufficient conditions that any PC matrix satisfies the POP and POIP conditions with respect to a vector generated by the GM method. The result is based on the well-known Koczkodaj's inconsistency index, KII, see [58].

Definition 2.6 *Koczkodaj's inconsistency index, $KII(A)$, of $n \times n$ PC matrix $A = \{a_{ij}\}$ is defined as*

$$KII(A) = \max \left\{ 1 - \min \left\{ \frac{a_{ij}}{a_{ik} a_{kj}}, \frac{a_{ik} a_{kj}}{a_{ij}} \right\} | i, j, k \in \{1, \dots, n\} \right\}. \quad (2.42)$$

It is obvious from (2.42) that $0 \leq KII(A) < 1$.

Proposition 2.13 *Let $A = \{a_{ij}\}$ be a PC matrix, let $w = (w_1, \ldots, w_n)$ be a priority vector generated by the GM method (2.24) and let $i, j, k, l \in \{1, \ldots, n\}$.*
Set $K = \frac{1}{1-KII(A)}$.
If $a_{ij} > K$, then $w_i > w_j$.
Moreover, if $a_{kl} > K$, and $\frac{a_{ij}}{a_{kl}} > K^2$,
then $w_k > w_l$, and $\frac{w_i}{w_j} > \frac{w_k}{w_l}$.
    Hence, the POP and POIP conditions are satisfied for A with respect to priority vector w.

Proof The proof follows directly from Theorem 3 in [62].

Proposition 2.13 guarantees that, for A and w generated by GMM, the POP condition holds, if the corresponding matrix element is large enough. If, for example, $KII(A) = 0.5$, then for the POP condition to be met, the element has to be greater than $K = \frac{1}{1-0.5} = 2$. Moreover, the proposition guarantees that the FS condition is satisfied as well.

As it is demonstrated in the following Example 2.5 (a), for B and $w^{(B)}$ generated by GMM, the POP condition is not satisfied, whereas the FS condition is satisfied. Moreover, in Example 2.5 (b), D and $w^{(D)}$ generated by GMM, the POP condition is satisfied, whereas POIP condition fails to be satisfied.

Example 2.5 Consider the set of four alternatives $\mathscr{C} = \{c_1, c_2, c_3, c_4\}$, and the corresponding PC matrix:
    (a) Consider a PC matrix $B = \{b_{ij}\}$ as follows:

$$B = \begin{pmatrix} 1 & \frac{4}{3} & \frac{3}{2} & \frac{3}{2} \\ \frac{3}{4} & 1 & 2 & 2 \\ \frac{2}{3} & \frac{1}{2} & 1 & 1 \\ \frac{2}{3} & \frac{1}{2} & 1 & 1 \end{pmatrix}.$$

And the priority vector (obtained by GM method, additively normalized)

$$w^{(B)} = (0.317, 0.317, 0.183, 0.183). \tag{2.43}$$

Clearly, $w_1 = 0.317$ is the maximal weight and c_1 is the non-dominated alternative, hence the FS condition is met. Moreover, inconsistency of B is acceptable as $CR = 0.016 < 0.08$.
    Here, $b_{12} = \frac{4}{3} > 1$, however, $\frac{w_1}{w_2} = \frac{0.317}{0.317} = 1.000$, therefore the POP condition is not met.

(b) Consider a PC matrix D given as follows:

$$D = \begin{pmatrix} 1 & 1 & 2 & 2 \\ 1 & 1 & 3 & 2 \\ \frac{1}{2} & \frac{1}{3} & 1 & 2 \\ \frac{1}{2} & \frac{1}{2} & \frac{1}{2} & 1 \end{pmatrix},$$

and the priority vector (obtained by the GM method, multiplicatively normalized)

$$w^{(D)} = (1.414, 1.565, 0.760, 0.595). \tag{2.44}$$

We obtain that $w_2 = 1.565$ is the maximal weight and c_2 is non-dominated alternative, hence the FS condition is met. Moreover, inconsistency of D is acceptable as $CR = 0.044 < 0.08$.

As it can be easily demonstrated, POP condition is satisfied:

$$d_{13} = 2, \frac{w_1}{w_3} = \frac{1.414}{0.760} > 1.000, d_{14} = 2, \frac{w_1}{w_4} = \frac{1.414}{0.595} > 1.000,$$

$$d_{23} = 3, \frac{w_2}{w_3} = \frac{1.565}{0.760} > 1.000, d_{24} = 2, \frac{w_2}{w_4} = \frac{1.565}{0.595} > 1.000,$$

$$c_{34} = 2, \frac{w_3}{w_4} = \frac{0.760}{0.595} > 1.000.$$

On the other hand, the POIP condition is not met, e.g.,

$$d_{23} = 3 > d_{14} = 2, \text{ and } \frac{w_2}{w_3} = \frac{1.565}{0.760} = 2.060 < \frac{w_1}{w_4} = \frac{1.414}{0.595} = 2.378.$$

Remark 2.4 The properties investigated here are formulated in terms of the usual "multiplicative system" of computing with the PC matrix elements. However, this system can be generalized to Abelian linearly ordered groups (alo-groups), see below or [18, 62].

It is highly desirable that for a given PC matrix, A, possibly inconsistent, we are able to generate a priority vector w such that the FS, POP, and POIP conditions are satisfied. For this purpose, we shall formulate a special optimization problem whose solution will generate the desirable priority vector associated with the PC matrix A satisfying all three above stated conditions.

2.7 Alternative Approach to Derivation of the Priority Vector

2.7.1 (Problem 0)

It was shown in Sect. 2.6, Example 2.4, that the calculation of a priority vector by the EV or GM methods from an inconsistent pairwise comparison matrix may result in violating the desirable conditions FS, POP, or POIP. Therefore, an alternative approach to the derivation of a priority vector for PCMs may be formulated in terms of satisfying the FS, POP, and POIP conditions.

Let $A = \{a_{ij}\}$ be a PC matrix. Based on this PCM, we need the following two sets of indexes:

$$I^{(2)}(A) = \{(i, j) | i, j \in \{1, \ldots, n\}, a_{ij} > 1\}, \tag{2.45}$$

$$I^{(4)}(A) = \{(i, j, k, l) | i, j, k, l \in \{1, \ldots, n\}, a_{ij} > 1, a_{kl} > 1, a_{ij} > a_{kl}\}. \tag{2.46}$$

Let

$$\delta : (x, y) \in \mathbf{R}_+ \times \mathbf{R}_+ \longrightarrow \delta(x, y) \in \mathbf{R}_+$$

be a *distance function*, i.e., a function with the following properties for all $x, y, z \in \mathbf{R}_+$:

- $\delta(x, y) \geq 0$,
- $\delta(x, y) = 0 \Leftrightarrow x = y$,
- $\delta(x, y) = \delta(y, x)$,
- $\delta(x, z) \leq \delta(x, y) + \delta(y, z)$.

Let $w = (w_1, \cdots, w_n)$ be a priority vector associated with A. An $n \times n$ *matrix of distances*, $\Delta(A, w)$, is defined as

$$\Delta(A, w) = \{\Delta_{ij}\} = \{\delta(a_{ij}, \frac{w_i}{w_j})\},$$

and a *matrix aggregation function*, Φ:

$$\Phi : X \in \mathbf{R}_+^n \times \mathbf{R}_+^n \longrightarrow \Phi(X) \in \mathbf{R}_+,$$

as an idempotent and increasing function (in each variable), where $X = \{x_{ij}\}$ is an $n \times n$ PC matrix.

An *error function*, $\mathscr{F}_A$, of $w = (w_1, \cdots, w_n)$ is defined as follows:

$$\mathscr{F}_A : w \in \mathbf{R}_+^n \longrightarrow \mathscr{F}_A(w) \in \mathbf{R}_+,$$

$$\mathscr{F}_A(w) = \Phi(\Delta(A, w)). \tag{2.47}$$

The problem of finding a priority vector satisfying the FS, POP, and POIP conditions can be formulated in terms of the following optimization problem, where $A = \{a_{ij}\}$ is a given PC matrix and $w = (w_1, \cdots, w_n)$ is an unknown priority vector with variables $w_1, \cdots, w_n$:

(Problem 0)

$$\mathscr{F}_A(w) \longrightarrow \min; \tag{2.48}$$

subject to

$$\prod_{r=1}^{n} w_r = 1, w_r > 0 \ \forall r, \tag{2.49}$$

$$w_r > w_s \ \forall (r, s) \in I^{(2)}(A), \tag{2.50}$$

$$\frac{w_r}{w_s} > \frac{w_t}{w_u} \ \forall (r, s, t, u) \in I^{(4)}(A). \tag{2.51}$$

The objective function in (2.48) minimizes the distance between the elements of PC matrix A and corresponding elements of the PCM ratio $W = \{\frac{w_i}{w_j}\}$, measured by distance function δ. By constraint (2.49), the weights are positive and (multiplicatively) normalized. By (3.59), the POP condition is secured and by (2.51) the POIP condition is satisfied.

2.7.2 Transformation to (Problem ε)

Unfortunately, (Problem 0) is not in the form of a standard optimization problem that is appropriate for solving by existing numerical methods. Here, variables w_i are required to be strictly positive and some inequality constraints, (2.50), (2.51), are strict, hence the set of feasible solution is not closed. That is why we transform the problem into a more convenient form. Given $\varepsilon > 0$.

(Problem ε)

$$\mathscr{F}_A(w) \longrightarrow \min; \tag{2.52}$$

subject to

$$\sum_{r=1}^{n} w_r = 1, w_r \geq \varepsilon \ \forall r, \tag{2.53}$$

$$w_r - w_s \geq \varepsilon \ \forall (r, s) \in I^{(2)}(A), \tag{2.54}$$

$$\frac{w_r}{w_s} - \frac{w_t}{w_u} \geq \varepsilon \ \forall (r, s, t, u) \in I^{(4)}(A). \tag{2.55}$$

In (Problem ε), nonlinear constraint (2.49) (with the product) is substituted by a linear constraint (2.53) (with the sum). Such a transform is possible as the multiplicative and additive normalization forms of priority vectors are equivalent (see e.g., Proposition 2.2).

Notice that here, strict inequalities have been changed to the non-strict ones by adding a sufficiently small constant $\varepsilon > 0$.

The following proposition says that both problems, i.e., (Problem 0) and (Problem ε), are in some sense equivalent. The proof of the proposition is evident.

Proposition 2.14 *(Problem 0) has a feasible solution w if and only if there exists $\varepsilon > 0$ such that w is a feasible solution of (Problem ε).*

Moreover, if (Problem 0) has an optimal solution w^* then there exists $\varepsilon > 0$ such that w^* is an optimal solution of (Problem ε).

Here, by (Problem ε) we denote the following three optimization problems depending on the particular formulation of the objective function (2.52) as well as constraints (2.53)–(2.55), i.e., nested sets of feasible solutions. Some examples are presented below. We shall consider the following optimization problem variants:

(I) Minimize the objective function (2.52), subject to (2.53). The optimal solution is denoted by $w^{(I)}$. The FS, POP, and POIP conditions are not necessarily satisfied.

(II) Minimize the objective function (2.52) subject to constraints (2.53), (2.54). The optimal solution is denoted by $w^{(II)}$. The POP condition is satisfied; then by Proposition 2.14 the FS condition is also satisfied. The POIP condition is not necessarily satisfied.

(III) Minimize the objective function (2.52) subject to constraints (2.53)–(2.55). The optimal solution is denoted by $w^{(III)}$. Here, the FS, POP, and POIP conditions are satisfied.

2.7.3   Solving (Problem ε)

Notice that the set of feasible solutions of (Problem ε), (2.52)–(2.55), could be empty, e.g., for problems (II), and/or (III), see below. Even for a nonempty set of feasible solutions of (Problem 0), the optimal solution of the corresponding optimization problems (I), (II), or (III) need not exist, as the set of feasible solutions is not secured to be closed and/or bounded and the objective function need not be convex.

On the other hand, if the optimal solution $w^* = (w_1^*, \cdots, w_n^*)$ of some problems of (I)–(III) of (Problem ε) exists, the FS, POP, RP, and POIP conditions hold by the nested properties of the feasible solution sets. Then, $w^* = (w_1^*, \cdots, w_n^*)$ is an appropriate priority vector associated with A satisfying the required properties.

Proposition 2.15 *Let $A = \{a_{ij}\}$ be a consistent pairwise comparison matrix. Then there is a unique optimal solution $w^* = (w_1^*, \cdots, w_n^*)$ of (Problem 0) satisfying:*

$$a_{ij} = \frac{w_i^*}{w_j^*} \ \forall i, j, \tag{2.56}$$

such that the FS, POP, RP, and POIP conditions are met.

Proof By Proposition 2.14 there exists a unique priority vector with positive components $w^* = (w_1^*, \cdots, w_n^*)$, such that

$$a_{ij} = \frac{w_i^*}{w_j^*} \quad \forall i, j. \tag{2.57}$$

Then by property (ii) of the distance function δ, it follows that $\delta(A, w^*) = \mathbf{0}$, i.e., $\delta(A, w^*)$ is a zero $n \times n$ matrix. As the matrix aggregation function Φ is idempotent, it holds the $\Phi(0) = 0$, which is the minimal value of (2.52). Hence, $w^* = (w_1^*, \cdots, w_n^*)$ is an optimal solution of (Problem 0). By (2.56) we have that the FS, POP, RP, and POIP conditions are met.   □

Now, we present some examples of simple distance functions $\delta(x, y)$ and the matrix aggregation function $\Phi(X)$.

Examples of distance functions $\delta(x, y)$: Let $x, y \in \mathbf{R}_+$.

i. $\delta(x, y) = |x - y|$,
ii. $\delta(x, y) = (x - y)^2$,
iii. $\delta(x, z) = (\ln(x) - \ln(y))^2$,
iv. $\delta(x, y) = \max\{\frac{x}{y}, \frac{y}{x}\}$.

Examples of aggregation functions $\Phi(X)$: Let $X = \{x_{ij}\}$ be a $n \times n$ matrix, $x_{ij} \in \mathbf{R}_+$.

a. $\Phi(X) = \frac{1}{n^2} \sum_{i,j=1}^{n} x_{ij}$,
b. $\Phi(X) = \max\{x_{ij} | i, j \in \{1, ..., n\}\}$.

Then the objective function in (Problem 0) and (Problem ε) is defined as

$$\mathscr{F}_A(w) = \Phi(\{\delta(a_{ij}, \frac{w_i}{w_j})\}). \tag{2.58}$$

In the sequel, we shall deal with the following particular items of the above examples: iv. and b. of metric function δ and aggregation function Φ, respectively, i.e.,

$$\delta(x, y) = \max\{\frac{x}{y}, \frac{y}{x}\}, \text{ and } \Phi(X) = \max\{x_{ij} | i, j \in \{1, ..., n\}\}. \tag{2.59}$$

Then we obtain the following objective function

$$\mathscr{F}(A, w) = \max\{\max\{\frac{a_{ij} w_j}{w_i}, \frac{w_i}{a_{ij} w_j}\} | i, j \in \{1, ..., n\}\}. \tag{2.60}$$

For other combinations of functions δ and Φ our approach presented below needs some modifications. When solving a particular optimization problem, (Problem ε), (2.52)–(2.55) with the objective function (2.60), we can encounter numerical difficulties, as this optimization problem is nonlinear and also nonconvex. Nonconvexity is found in objective function (2.60) and also in constraints (2.55). Fortunately, these obstacles can be avoided by a proper approach—transformation of the nonconvex problem to a convex one, which enables using standard numerical methods for solving NLP problems. Then, for variants (I) and (II) of (Problem ε), we obtain an optimization problem solvable, e.g., by efficient interior point methods (see e.g., [13]). For solving variant (III) with nonconvex constraints (2.55), we can apply, e.g., an interior or exterior penalty method by penalizing this constraint and moving it into the objective function (see e.g., [13] or [76]).

First, we analyze the objective function (2.60). Here, setting

$$f_{ij}(w) = \frac{a_{ij}w_i}{w_j}, i, j \in \{1, \cdots, n\},\tag{2.61}$$

where $w = (w_1, \cdots, w_n)$, we obtain a simplified form of the linear fractional function on $\mathbf{R}^n_+$, which is a quotient of two linear functions. Function (2.61) is not convex, it is, however, quasiconvex. More precisely, it is strictly quasiconvex on $\mathbf{R}^n_+$, the positive orthant of $\mathbf{R}^n$.

Recall that a function $f_{ij} : \mathbf{R}^n_+ \to \mathbf{R}$, is *strictly quasiconvex on* $\mathbf{R}^n_+$ if

$$f_{ij}(\lambda x + (1 - \lambda)y) < \max\{f_{ij}(x), f_{ij}(y)\}$$

holds for all $\lambda \in]0; 1[$ and all $x, y \in \mathbf{R}^n_+$. Moreover, function $f_{ij} : \mathbf{R}^n_+ \to \mathbf{R}$ is *strictly quasiconcave on* $\mathbf{R}^n_+$ if $(-f_{ij})$ is strictly quasiconvex on $\mathbf{R}^n_+$. At the same time function (2.60) is strictly quasiconcave on $\mathbf{R}^n_+$. Hence, the reciprocal function

$$\frac{1}{f_{ij}(w)} = \frac{w_j}{a_{ij}w_i}$$

is strictly quasiconvex on $\mathbf{R}^n_+$. The pointwise maximum of two strictly quasiconvex functions on $\mathbf{R}^n_+$ is again strictly quasiconvex on $\mathbf{R}^n_+$,

$$g_{ij}(w) = max\{\frac{a_{ij}w_i}{w_j}, \frac{w_j}{a_{ij}w_i}\}$$

is strictly quasiconvex on $\mathbf{R}^n_+$. Moreover, the pointwise maximum over all functions $g_{ij}, i, j \in \{1, ..., n\}$ is strictly quasiconvex on $\mathbf{R}^n_+$, and

$$\mathscr{F}(A, w) = \max\{\max\{\frac{a_{ij}w_j}{w_i}, \frac{w_i}{a_{ij}w_j}\}|i, j \in \{1, ..., n\}\}$$

is strictly quasiconvex on $\mathbf{R}^n_+$.

It is a well-known fact says that strictly quasiconvex functions are unimodal, i.e., each local minimum of each strictly quasiconvex function is a global minimum (see e.g., [13]). Summarizing the above stated results, we obtain that the objective function (2.59) is unimodal. Taking into account the fact that constraints (2.52),

(2.54) in (Problem ε), i.e., variant (II), define a convex set, we conclude that the set of all optimal solutions of (Problem ε), variant (II), is convex and each local optimal solution is global. Consequently, by solving (Problem ε), variant (II), e.g., by some interior point method (see [13]), we arrive at the global optimal solution.

Alternatively, variant (II) of (Problem ε) can be solved by a sequence of linear problems as follows (epigraph method):

Instead of minimizing objective (2.59) subject to constraints (2.53), (2.54), we set $t = \frac{a_{ij} w_i}{w_j}$ and solve the following system of linear constraints in each iteration:

$$a_{ij} w_i - w_j t \leq 0, \text{ for all } i, j, \tag{2.62}$$

$$w_j - a_{ij} w_i t \leq 0, \text{ for all } i, j,$$

$$\sum_{r=1}^{n} w_r = 1, w_r \geq \varepsilon \text{ for all } r,$$

$$w_r - w_s \geq \varepsilon \text{ for all } (r, s) \in I^{(2)}(A),$$

then adapting t in each step by the well-known bisection optimization method, [13]. The solutions of system (2.62) generate a sequence that converges to the optimal solution of variant (II) of (Problem ε).

Constraints (2.55) in (Problem ε), however, need a special treatment. The set of vectors $w = (w_1, \cdots, w_n)$ fulfilling constraints (2.55) is neither convex nor star-shaped (see [78] or [90]), and therefore, the usual interior point methods for solving the optimization problem (2.59), (2.53)–(2.55) could be inefficient or fail. That is why we propose solving (Problem ε), variant (III), the popular penalty methods, see e.g., [13], or [76]. The main idea of penalty methods is to penalize difficult constraints of the problem and move them to the objective function. Then the remaining constraints of (Problem ε) are fairly tractable. They are linear, so the problem can be solved more easily by interior point methods, for example. The form of the constraints (2.55) is suitable for the application of the interior penalty methods (barrier methods), as it satisfies some appropriate assumptions: an interior point of the constraint set can be easily found, the values of the constraint functions are unlimited when approaching from inside to the border of $\mathbf{R}_{+,,}^n$, etc.

2.7.4 Illustrative Example

In this section, we present an illustrative example of the 4×4 PC matrix from Example 2.1. All results have been obtained by Excel Solver.

Example 2.6 Consider the set of four alternatives $n = 4$, $\mathscr{C} = \{c_1, c_2, c_3, c_4\}$, and the corresponding PC matrix A from Example 2.1 given as follows:

$$A = \{a_{ij}\} = \begin{pmatrix} 1 & \frac{3}{2} & 2 & 2 \\ \frac{2}{3} & 1 & 4 & 4 \\ \frac{1}{2} & \frac{1}{4} & 1 & 1 \\ \frac{1}{2} & \frac{1}{4} & 1 & 1 \end{pmatrix}.$$

From the first row of PC matrix A, the alternative c_1 clearly dominates the other three alternatives, hence, c_1 is non-dominated. Saaty's consistency index CI and consistency ratio CR are: $CI = 0.052$, $RI = 0.089$, $CR = CI/RI = 0,058$. Hence, by Saaty, inconsistency is acceptable and the priority vector generated by the EVM (additively normalized) is as follows: $w_{EV} = (0.350, 0.396, 0.127, 0.127)$.

This priority vector corresponds to the ranking of the four alternatives $\mathscr{C} = \{c_1, c_2, c_3, c_4\}$ as follows:

$$c_2 \succ c_1 \succ c_3 \simeq c_4.$$

The priority vector generated by the GMM is $w_{GM} = (0.343, 0.396, 0.130, 0.130)$.

This priority vector corresponds to the same ranking of alternatives $\mathscr{C} = \{c_1, c_2, c_3, c_4\}$ as follows:

$$c_2 \succ c_1 \succ c_3 \simeq c_4.$$

Set $\varepsilon = 0.0001$. We solve (Problem ε) as follows:

We consider the following items of the above examples: iv. and b. of metric function δ and aggregation function Φ, respectively,

$$\delta(x, y) = \max\{\frac{x}{y}, \frac{y}{x}\}, \text{ and } \Phi(X) = \max\{x_{ij} | i, j \in \{1, ..., 4\}\}.$$

We obtain

$$\mathscr{F}(A, w) = \max\{\max\{\frac{a_{ij}w_j}{w_i}, \frac{w_i}{a_{ij}w_j}\} | i, j \in \{1, 2, 3, 4\}\}.$$

In variant (I) of (Problem ε), we shall minimize objective function (2.52) subject to constraint (2.53).

Then, solving (Problem ε), we obtain the optimal solution

$$w^{(Ib)} = (0.378, 0.361, 0.131, 0.131)$$

with the corresponding minimal error function: $\mathscr{F}(A, w^{(Ib)}) = 1.448$.

This priority vector corresponds to the ranking of our alternatives $\mathscr{C} = \{c_1, c_2, c_3, c_4\}$ as follows:

$$c_1 \succ c_2 \succ c_3 \simeq c_4.$$

Here, the POP and FS conditions are satisfied, as can be immediately verified.

We obtain $a_{12} = \frac{3}{2} > 1$, and, $\frac{w_1}{w_2} = \frac{0.378}{0.361} > 1.000$. Similarly, $a_{13} = a_{14} = 2 > 1$, and, $\frac{w_1}{w_3} = \frac{w_1}{w_4} = \frac{0.378}{0.131} > 1.000$, and, $a_{23} = a_{24} = 4 > 1$, and, $\frac{w_2}{w_3} = \frac{w_2}{w_4} = \frac{0.361}{0.131} > 1.000$. Therefore, the POP condition is met.

Moreover, the POIP condition is not satisfied, e.g.,

$$a_{23} = 4 > a_{13} = 2, \text{ and } \frac{w_2}{w_3} = \frac{0.361}{0.131} < \frac{w_1}{w_3} = \frac{0.378}{0.131}.$$

In variant (II) the objective function (2.52) is minimized subject to constraints (2.53), (2.54). Evidently, here, we obtain the same optimal solution as in Variant (I), i.e.

$$w^{(Ib)} = (0.378, 0.361, 0.131, 0.131).$$

By the same arguments as in variant (I), the POP and FS conditions are satisfied.

In variant (III) the objective function (2.52) is minimized subject to constraints (2.53)–(2.55). The POP and POIP conditions should be satisfied simultaneously, as well as the FS condition. By solving (Problem ε), we obtain that the set of feasible solutions (2.53)–(2.55) is empty. On the other hand, by omitting constraints (2.54), i.e., the POP condition, we obtain the following optimal solution:

$$w^{(IIIb)} = (0.378, 0.361, 0.131, 0.131) = w^{(Ib)}.$$

Here, the POP as well as FS conditions are not met.

2.8 Additive Pairwise Comparison Matrices

The abovementioned interpretation of the preference relations on the set of alternatives $\mathscr{C} = \{c_1, c_2, \cdots, c_n\}$ described by a multiplicative pairwise comparison matrix $A = \{a_{ij}\}$ (e.g., in AHP, see e.g., [92]) is, however, not always appropriate for a decision-maker. When evaluating the preference of two elements of a pair, say, c_i and c_j with respect to, e.g., "design of products," or "hight of person," it might cause a problem. Here, saying, for example, that alternative c_i is 3 times better than c_j, or, particularly, that Paul is 0.25 times taller than Peter, might be peculiar. It appears that Saaty s ratio scale for pairwise comparisons does not represent reality and often produces an exaggerated results contrary to common sense, although mathematically the operation of the m-reciprocal matrix seems to be useful. The perception of the interval, or difference, between two alternatives is relatively much simpler than the perception of the ratio of both.

In this section, the preferences over the set of alternatives X will be represented in the following way. Let us assume that the intensities of the DM's preferences are given by an $n \times n$ matrix $A = \{a_{ij}\}$ with real number elements in such a way that, for all i and j, the entry a_{ij} indicates the difference of preference intensity for alternative xc_i to that of c_j. In other words, a_{ij} means that the difference between intensities c_i

and c_j is a_{ij}. Hence, element a_{ij} can be positive, zero or negative. The perception of the interval (or difference) between two objects is relatively much simpler than the perception of the ratio of both. The reasons are that the operations of addition and subtraction are easier than the operations of multiplication and division, which, in fact, are based on repeated addition and subtraction, respectively. Therefore, addition and subtraction are straightforward for the comparison of two objects.

K.T.F. Yuen in [11] suggests representing the preference intensities by the relative scale $\{-\kappa, -7\kappa/8, -6\kappa/8, \cdots, -\kappa/8, 0, \kappa/8, 2\kappa/8, \cdots, 7\kappa/8, \kappa\}$ in contrast to the Saatys absolute scale $\{1/9, 1/8, \cdots, 1/2, 1, 2, \cdots, 8, 9\}$.

Here, $a_{ij} = 0$ indicates the equal intensity of both alternatives and $a_{ij} = \kappa$, is the maximal difference (i.e., utility) of the intensities between the compared alternatives, [11]. Here, κ indicates how people perceive the difference between paired objects in different scenarios. Here, $2 \times 8 + 1 = 17$ is the number of the intervals of the scale schema which corresponds to 7 ± 2 stages in a single rating process. In general, the scale for measuring preference intensities is the interval of all real numbers, i.e., $\mathbf{R} =]-\infty, +\infty[$.

The elements of $A = \{a_{ij}\}$ satisfy the following reciprocity condition. A real $n \times n$ matrix $A = \{a_{ij}\}$ is *additive-reciprocal* (*a-reciprocal*) if

$$a_{ji} = -a_{ij} \text{ for all } i, j \in \{1, \cdots, n\} . \tag{2.63}$$

A real $n \times n$ matrix $A = \{a_{ij}\}$ is *additive-consistent* (or, *a-consistent*) [44, 93], if

$$a_{ik} = a_{ij} + a_{jk} \text{ for all } i, j, k \in \{1, \cdots, n\} . \tag{2.64}$$

From (2.64) we obtain $a_{ii} = 0$ for all i, and also (2.64) implies (2.63), i.e., an a-consistent matrix is a-reciprocal (but not vice-versa).

Here, all matrices are $n \times n$-matrices with real number elements and all vectors are n-dimensional vectors with real number elements.

Let $\mathscr{A}^+$ be the set of all a-reciprocal matrices. Then $\mathscr{A}^+$ is an additive group under componentwise addition. If $A = \{a_{ij}\} \in \mathscr{A}^+$ and $B = \{b_{ij}\} \in \mathscr{A}^+$, then $C = A + B = \{a_{ij} + b_{ij}\} \in \mathscr{A}^+$. Here, by "+" we denote the group operation, i.e., the componentwise addition of matrices. Similarly, let $\mathscr{A}^{+c}$ be the set of all a-consistent matrices and v^+ be the set of all vectors $v = (v_1, v_2, \cdots, v_n)$ with real number elements such that $\sum_{j=1}^{n} v_j = 0$. Both $\mathscr{A}^{+c}$ and v^+ are additive groups under componentwise addition. Moreover, $\mathscr{A}^{+c}$ is a subgroup of $\mathscr{A}^+$ and, as we will see by Proposition 2.16, $\mathscr{A}^{+c}$ is isomorphic to v^+.

Let $\mathscr{F}^+$ be the set of all mappings from $\mathscr{A}^+$ into v^+, with $f \in \mathscr{F}^+$. Vector $v = (v_1, v_2, \cdots, v_n) \in v^+$ is called the *a-priority vector of* $A = \{a_{ij}\}$, or the a-priority vector associated with $A = \{a_{ij}\}$, if there exists $f \in \mathscr{F}^+$ such that $v = f(A)$.

If $v = (v_1, v_2, \cdots, v_n)$ is the a-priority vector of a-consistent matrix $A = \{a_{ij}\}$, then the following condition holds:

$$a_{ij} = v_i - v_j \text{ for all } i, j \in \{1, \cdots ., n\} , \tag{2.65}$$

and v is called the *a-consistency vector of A*.

Here, the mapping f defines how the a-priority vector is "calculated" from the elements a_{ij} of A.

Proposition 2.16 *A matrix $A = \{a_{ij}\}$ is a-consistent if and only if there exists a vector $v = (v_1, v_2, \cdots, v_n)$ with $\sum_{j=1}^{n} v_j = 0$ such that*

$$a_{ij} = v_i - v_j \text{ for all } i, j \in \{1, \cdots, n\} . \tag{2.66}$$

Proof (i) Let let $A = \{a_{ij}\}$ be a-consistent, then by setting

$$v_i = \frac{1}{n} \sum_{k=1}^{n} a_{ik} \text{ for all } i \in \{1, \cdots, n\} ,$$

we obtain

$$v_i - v_j = \frac{1}{n} \sum_{k=1}^{n} (a_{ik} - a_{jk}) = a_{ij} \text{ for all } i \in \{1, \cdots, n\} .$$

Considering a-reciprocity (2.64) we obtain

$$\sum_{i=1}^{n} v_i = \frac{1}{n} \sum_{i=1}^{n} \sum_{k=1}^{n} a_{ik} = 0 \text{ for all } i \in \{1, \cdots, n\} .$$

(ii) Let (2.66) be true, then (2.63) is easily satisfied.   □

In the DM problem given by $\mathscr{C} = \{c_1, c_2, \cdots, c_n\}$ and $A = \{a_{ij}\}$, the rating of the alternatives in $\mathscr{C}$ is determined by the the a-priority vector $v = (v_1, v_2, \cdots, v_n)$ of A. The ranking of alternatives is defined as follows:

$$c_i \succ c_j \text{ if } v_i > v_j \text{ for all } i, j \in \{1, \cdots, n\} .$$

Therefore, the alternatives $c_1, c_2, \cdots, c_n$ in $\mathscr{C}$ can be ranked by the natural ranking of the components of the corresponding a-priority vector.

Generally, the PC matrix $A = \{a_{ij}\}$ is not a-consistent. Hence the a-priority vector cannot be an a-consistency vector. We face the problem of how to measure the inconsistency of the PC matrix. To solve this problem we shall define special a-consistency indexes based on the corresponding a-priority vectors.

2.8.1 Deriving Priority Vector from Additive PCM

The least squares method (LSM), presented in Sect. 2.5.3 makes use also of the additive properties of the PC matrix $A = \{a_{ij}\} \in \mathscr{A}^+$. Here, the LSM assumes the minimization of the sum of the squared deviations from given elements of A, i.e.,

$$\sum_{1 \leq i < j \leq n} \left(a_{ij} - (v_i - v_j) \right)^2 \longrightarrow \min; \tag{2.67}$$

subject to

$$\sum_{j=1}^{n} v_j = 0 . \tag{2.68}$$

As is demonstrated, for example, in [34] the optimal solution of problem (2.67), (2.68) can be found simply as the arithmetic mean of the rows of an a-reciprocal PC matrix. The elements of the a-priority vector—the optimal solution of problem (2.67), (2.68) $v^* = (v_1^*, v_2^*, \cdots, v_n^*)$ are defined as the arithmetic mean of the row elements of A:

$$v_i^* = \frac{1}{n} \sum_{j=1}^{n} a_{ij} \text{ for all } i \in \{1, \cdots, n\} . \tag{2.69}$$

An a-consistency measure of a matrix $A = \{a_{ij}\}$ with respect to LSM is given by the *a-consistency index of A*, $I_a(A)$, defined as the minimal value of the objective function (2.67):

$$I_{aLS}(A) = \sum_{1 \leq i < j \leq n} (e_{ij}^a)^2 , \tag{2.70}$$

where $v^* = (v_1^*, v_2^*, \cdots, v_n^*)$ is given by (2.69), the optimal solution of (2.67), (2.68), and moreover, $E_A^a = \{e_{ij}^a\} = \{a_{ij} - (v_i^* - v_j^*)\}$, $V_A^a = \{v_i^* - v_j^*\}$.

Here, $E_A^a = \{e_{ij}^a\}$ is called the *a-error matrix of* $A = \{a_{ij}\}$. By (2.69) we obtain $e_{ij}^a = \frac{1}{n} \sum_{k=1}^{n} (a_{ij} + a_{jk} + a_{ki})$.

Notice that $V_A^a = \{v_i^* - v_j^*\} \in \mathscr{A}^{+c}$, i.e., the matrix V_A^a is a-consistent. Then evidently

$$A = E_A^a - V_A^a . \tag{2.71}$$

The proof of the following proposition which is parallel to Proposition 2.7 is straightforward.

Proposition 2.17 *If $A = \{a_{ij}\}$ is an a-reciprocal matrix, then $I_{aLS}(A) \geq 0$. Moreover, A is a-consistent if and only if $I_{aLS}(A) = 0$.*

Proposition 2.17 enables us to distinguish between a-consistent and a-inconsistent matrices but it is insufficient to determine the degree of a-consistency from a-inconsistent matrices. In fact, a statement of the type: *A is less a-consistent than B if $I_{aLS}(A) > I_{aLS}(B)$* is not meaningful when A and B are of different dimensions. Moreover, a cut-off rule of the type: *A is close enough to being a-consistent if $I_{aLS}(A) \leq \varepsilon$* for some fixed positive constant ε, independent of the dimension n, does not appear to be meaningful.

We are going to construct a better measure than the above defined a-consistency index I_{aLS}. A natural way that would preserve the abovementioned properties and addresses its deficiencies is to consider the *relative a-error* $RE_a(A)$ *of* $A = \{a_{ij}\}$ defined as the "normalized" I_{aLS} (see [7]):

$$RE_a(A) = \frac{\sum_{1 \le i < j \le n} (e_{ij}^a)^2}{\sum_{1 \le i < j \le n} a_{ij}^2} . \tag{2.72}$$

Proposition 2.18 *If* $A = \{a_{ij}\}$ *is an a-reciprocal matrix, then*

$$0 \le RE_a(A) \le 1 . \tag{2.73}$$

Proof Since the error component minimizes the sum of squares (2.67) we have

$$\sum_{1 \le i < j \le n} (a_{ij} - (v_i^* - v_j^*))^2 \le \sum_{1 \le i < j \le n} (a_{ij} - (v_i - v_j))^2 \le \sum_{1 \le i < j \le n} a_{ij}^2 .$$

Hence, (2.73) is satisfied.   □

By Proposition 2.18 the values of the relative a-error belong to the unit interval [0; 1] regardless of n, the dimension of A. Using this measure, we may compare the consistency of matrices of different dimensions and justify the use of cut-off rules accepting A as sufficiently consistent if, for example, its relative a-error satisfies $RE_a(A) \le 0.1$.

A matrix A is called *totally a-inconsistent* if its relative a-error is maximal, i.e., $RE_a(A) = 1$.

The relative a-error measures the relative grade of a-inconsistency of A. Now, we define a new index which will measure the relative grade of a-consistency of a matrix A by the matrix V_A^a.

The *relative a-consistency index* $RC_a(A)$ of $A = \{a_{ij}\}$ is defined as follows, see [7]

$$RC_a(A) = \frac{\sum_{1 \le i < j \le n} (v_i^* - v_j^*)^2}{\sum_{1 \le i < j \le n} a_{ij}^2} , \tag{2.74}$$

where $v^* = (v_1^*, v_2^*, \cdots, v_n^*)$ is given by (2.69), the optimal solution of (2.67), (2.68).

The main result of this subsection is based on the classical projection theorem (see [45]), applied to PC matrix A understood as a vector.

Theorem 2.3 *Let A be a vector in $\mathbf{R}^N$, N be a positive integer, and S be a subspace of $\mathbf{R}^N$. The vector A can be uniquely represented in the form:*

$$A = C + E , \tag{2.75}$$

where $C \in S$ and $E \perp S$. Moreover, C is a unique vector satisfying

$$\|A - C\| \leq \|A - X\| \text{ for all } X \in S ,\tag{2.76}$$

where $\|A\|$ is the Euclidean norm of A.

Now, we formulate the main result:

Proposition 2.19 If $A = \{a_{ij}\}$ is an a-reciprocal matrix, then

$$RC_a(A) + RE_a(A) = 1 .\tag{2.77}$$

Proof We may rewrite the $n \times n$ matrix A as an $N = n^2$-dimensional vector by ordering its elements by rows. In the N-dimensional Euclidean space $\mathbf{R}^N$ the vectors corresponding to the a-consistent matrices form a subspace S. Combining the projection theorem with the solution of the minimization problem which is characterized by the arithmetic mean, we see that the decomposition of A into its a-consistent and a-inconsistent components is an orthogonal decomposition, i.e., $C \perp E$, or

$$\sum_{1 \leq i < j \leq n} c_{ij}^a e_{ij}^a = 0 .$$

Using this equation with $a_{ij} = c_{ij}^a + e_{ij}^a$, we obtain

$$\sum_{1 \leq i < j \leq n} a_{ij}^2 = \sum_{1 \leq i < j \leq n} (c_{ij}^a)^2 + 2 \sum_{1 \leq i < j \leq n} c_{ij}^a e_{ij}^a + \sum_{1 \leq i < j \leq n} (e_{ij}^a)^2 ,$$

and therefore

$$\sum_{1 \leq i < j \leq n} a_{ij}^2 = \sum_{1 \leq i < j \leq n} (c_{ij}^a)^2 + \sum_{1 \leq i < j \leq n} (e_{ij}^a)^2 .$$

Hence, for $A \neq 0$ we obtain (2.77). □

Now, we derive the parallel result to Proposition 2.10.

Proposition 2.20 $A = \{a_{ij}\} \in \mathscr{A}^+$ is totally a-inconsistent if and only if the row sums of A are all zeros, i.e.,

$$\sum_{j=1}^n a_{ij} = 0 \text{ for all } i \in \{1, \cdots, n\} .\tag{2.78}$$

Proof (i) Let $A = \{a_{ij}\}$ be totally a-inconsistent, i.e., $RE_a(A) = 1$. Hence, by Proposition 2.19 we obtain

$$RC_a(A) = \frac{\sum_{1 \leq i < j \leq n} (v_i^* - v_j^*)^2}{\sum_{1 \leq i < j \leq n} a_{ij}^2} = 0 .$$

Therefore, $v_i^* = v_j^*$ for all $i, j = 1, \cdots, n$ and by (2.68) we have $v_i^* = 0$ for all $i \in \{1, \cdots, n\}$. Then by (2.69) we obtain

$$\frac{1}{n} \sum_{j=1}^{n} a_{ij} = v_i^* = 0 \text{ for all } i \in \{1, \cdots, n\} \ .$$

Hence, (2.78) holds.

(ii) Let (2.78) be satisfied. Then by (2.69) $v_i^* = \frac{1}{n} \sum_{j=1}^{n} a_{ij} = 0$ for all $i, j \in \{1, \cdots, n\}$. Consequently, $e_{ij}^a = a_{ij}$ for all $i, j \in \{1, \cdots, n\}$ and

$$\sum_{1 \le i < j \le n} (e_{ij}^a)^2 = \sum_{1 \le i < j \le n} a_{ij}^2 \ .$$

Hence, $RE_a(A) = 1$, i.e., A is totally a-inconsistent. $\qquad\square$

We illustrate the theory by a simple example.

Example 2.7 Let $\mathscr{C} = \{c_1, c_2, c_3\}$ be the set of alternatives and assume that the intensities of preferences are given by an 3×3 matrix $A = \{a_{ij}\} \in \mathscr{A}^+$ defined as

$$A = \begin{pmatrix} 0 & 1 & -4 \\ -1 & 0 & 7 \\ 4 & -7 & 0 \end{pmatrix} \ .$$

We can now compute the priority vector v^* as follows:

$$v^* = (v_1^*, v_2^*, v_3^*) = (\frac{1}{3}(0+1-4), \frac{1}{3}(-1+0+7), \frac{1}{3}(4-7+0)) = (-1, 2, -1) \ .$$

Then

$$V_A^a = \{v_i^* - v_j^*\} = \begin{pmatrix} 0 & -3 & 0 \\ 3 & 0 & 3 \\ 0 & -3 & 0 \end{pmatrix} \ ,$$

$$RC_a(A) = \frac{\sum_{1 \le i < j \le 3} (v_i^* - v_j^*)^2}{\sum_{1 \le i < j \le 3} a_{ij}^2} = \frac{18}{66} = 0.27.$$

Moreover, by (2.31) we have $RE_a(A) = 1 - RC_a(A) = 1 - 0.27 = 0.73$. We conclude that the relative a-consistency of A is 27% and the relative a-error of A is 73%. Hence, the a-inconsistency of A is relatively high.

According to the priority vector v^* we obtain the ranking: $c_2 \succ c_3 = c_1$.

2.9 Fuzzy Pairwise Comparison Matrices

Sometimes it is more natural, when comparing c_i to c_j, that the decision-maker assigns the positive value b_{ij} to c_i and b_{ji} to c_j, where $b_{ij} + b_{ji} = 1$. Here, $b_{ij}, b_{ji} \in S$, the preferences on $\mathscr{C} = \{c_1, c_2, \cdots, c_n\}$ can be understood as a *valued relation*, or *fuzzy preference relation*, with membership function $\mu : \mathscr{C} \times \mathscr{C} \rightarrow [0, 1]$, and $\mu(c_i, c_j) = b_{ij}$ denotes the preference of the alternative c_i over c_j [44, 81, 93]. Therefore, the fuzzy preference relation on $\mathscr{C}$ can be represented as a *fuzzy pairwise comparison matrix (PCF matrix)*.

In comparison to the previous DM problem, here, the scale S is always the unit interval [0, 1], otherwise, the DM problem remains the same. This means, that the elements b_{ij} of pairwise comparisons matrix $B = \{b_{ij}\}$ are crisp values, as before. For the sake of distinctions, we denote this PC matrix by symbols b_{ij}, instead of a_{ij}, with the shortcut of PCF, instead of PC. On the other hand, in Chap. 4, we shall deal with PC matrices with *fuzzy elements* and we shall denote such matrices with the shortcut FPC.

Here, $b_{ij} = \frac{1}{2}$ indicates indifference between c_i and c_j, $b_{ij} > \frac{1}{2}$ indicates that c_i is preferred to c_j, $b_{ij} = 1$ indicates that c_i is absolutely preferred to c_j, $b_{ij} = 0$ indicates that c_j is absolutely preferred to c_i. Other important properties of fuzzy pairwise comparison matrix $B = \{b_{ij}\}$ can be presented as follows.

An $n \times n$ FPC matrix $B = \{b_{ij}\}$ with $0 \leq b_{ij} \leq 1$ for all $i, j \in \{1, 2, ..., n\}$ is *fuzzy-reciprocal (f-reciprocal)* [29], if

$$b_{ij} + b_{ji} = 1 \text{ for all } i, j \in \{1, \cdots, n\} , \qquad (2.79)$$

or, equivalently

$$b_{ji} = 1 - b_{ij} \text{ for all } i, j \in \{1, \cdots, n\} . \qquad (2.80)$$

Evidently, if (2.79) holds, then $b_{ii} = \frac{1}{2}$ for all $i \in \{1, \cdots, n\}$.

For making a "coherent" choice (when assuming PCF matrices) some properties have been suggested in the literature [44, 102].

The nomenclature of properties of fuzzy PC matrices (i.e., FPCMs) has, however, not been stabilized yet, compare, e.g., [29, 71, 102, 106]. Here, we use the usual nomenclature which is as close as possible to the one used in the literature.

We say that PCF matrix $B = \{b_{ij}\}$ is *fuzzy-additive-consistent (fa-consistent)* [29], if

$$b_{ik} - \frac{1}{2} = (b_{ij} - \frac{1}{2}) + (b_{jk} - \frac{1}{2}) \text{ for all } i, j, k \in \{1, \cdots, n\} . \qquad (2.81)$$

Equation (2.81) can be equivalently rewritten as (see [102])

$$b_{ik} = \frac{1}{2} + b_{ij} - b_{kj} \text{ for all } i, j, k \in \{1, \cdots, n\} . \qquad (2.82)$$

or, see [29],

$$b_{ij} + b_{jk} + b_{ki} = \frac{3}{2} \text{ for all } i, j, k \in \{1, \cdots, n\} . \tag{2.83}$$

Notice that if $B = \{b_{ij}\}$ is fa-consistent, then B is f-reciprocal.

Let FPC matrix $B = \{b_{ij}\}$ be fuzzy-reciprocal according to (2.79). We say that $B = \{b_{ij}\}$ is *fuzzy-multiplicative-consistent (fm-consistent)*, [20] if

$$\frac{b_{ik}}{b_{ki}} = \frac{b_{ij}}{b_{ji}} \cdot \frac{b_{jk}}{b_{kj}} \text{ for all } i, j, k \in \{1, \cdots, n\}. \tag{2.84}$$

This property is sometimes called "transitivity" (see e.g., [71]); however, here we reserve this name for a different concept (see below).

It is not difficult to prove some relationships among individual consistency properties of FPC matrices.

Finally, consider the following 4 simple examples. (1) Let $A_1 = \{a_{1ij}\}$ be defined as

$$A_1 = \begin{pmatrix} 0.5 & 0.6 & 0.5 \\ 0.4 & 0.5 & 0.4 \\ 0.5 & 0.6 & 0.5 \end{pmatrix} .$$

Here, A_1 is f-reciprocal, fm-consistent, and fa-consistent. (2) Let $A_2 = \{a_{2ij}\}$ be defined as

$$A_2 = \begin{pmatrix} 0.5 & 0.3 & 0.2 \\ 0.7 & 0.5 & 0.4 \\ 0.8 & 0.6 & 0.5 \end{pmatrix} .$$

Here, A_2 is fa-consistent and NOT fm-consistent.

(3) Let $A_3 = \{a_{3ij}\}$ be defined as

$$A_3 = \begin{pmatrix} 0.5 & 0.25 & 0.1 \\ 0.75 & 0.5 & 0.25 \\ 0.9 & 0.75 & 0.5 \end{pmatrix} .$$

Here, A_3 is fm-consistent and NOT fa-consistent. (4) Let $A_4 = \{a_{4ij}\}$ be defined as

$$A_4 = \begin{pmatrix} 0.5 & 0.9 & 0.2 \\ 0.9 & 0.5 & 0.3 \\ 0.2 & 0.3 & 0.5 \end{pmatrix} .$$

Here, A_4 is fm-consistent and NOT f-reciprocal.

2.9.1 Some Relations Between Fuzzy Pairwise Comparison Matrices

In what follows, we shall investigate some relationships between f-reciprocal and m-reciprocal pairwise comparison matrices. We start with an extension of the result published by E. Herrera-Viedma et al. [29]. For this purpose, given $\sigma > 1$, we define the following function φ_σ and its inverse function φ_σ^{-1} as

$$\varphi_\sigma(t) = \frac{1}{2}(1 + \frac{\ln t}{\ln \sigma}) \text{ for } t \in [\frac{1}{\sigma}, \sigma] , \tag{2.85}$$

$$\varphi_\sigma^{-1}(t) = \sigma^{2t-1} \text{ for } t \in [0, 1] . \tag{2.86}$$

We obtain the following results, characterizing fa-consistent and m-consistent matrices, see [29, 81]. By $\sigma > 1$ evaluation scale $[1/\sigma; \sigma]$ is defined. In [29], only $\sigma = 9$ is investigated.

Proposition 2.21 *Let* $\sigma > 1$, $A = \{a_{ij}\}$ *be an* $n \times n$ *matrix with* $\frac{1}{\sigma} \le a_{ij} \le \sigma$ *for all* $i, j \in \{1, 2, ..., n\}$.
  If $A = \{a_{ij}\}$ *is m-consistent then* $B = \{\varphi_\sigma(a_{ij})\}$ *is fa-consistent.*

Proof Let let $A = \{a_{ij}\}$ be an $n \times n$ matrix with $\frac{1}{\sigma} \le a_{ij} \le \sigma$ for all i and j. Suppose that A is m-consistent and set $b_{ij} = \varphi_\sigma(a_{ij})$ for all i and j. Then by (2.83) and (2.85)

$$b_{ij} + b_{jk} + b_{ki} = \frac{1}{2}(3 + \frac{\ln a_{ij}a_{jk}a_{ki}}{\ln \sigma}) \text{ for all } i, j, k \in \{1 \cdots, n\} .$$

Hence, by (2.3) we have $a_{ij}a_{jk}a_{ki} = 1$ for all $i, j, k \in \{1, \cdots, n\}$ and then

$$b_{ij} + b_{jk} + b_{ki} = \frac{3}{2} \text{ for all } i, j, k \in \{1, \cdots, n\} .$$

and $B = \{\varphi_\sigma(a_{ij})\}$ is fa-consistent.   □

Proposition 2.22 *Let* $\sigma > 1$, $B = \{b_{ij}\}$ *be an* $n \times n$ *matrix with* $0 \le b_{ij} \le 1$ *for all* $i, j \in \{1, \cdots, n\}$.
  If $B = \{b_{ij}\}$ *is fa-consistent then* $A = \{\varphi_\sigma^{-1}(b_{ij})\}$ *is m-consistent.*

Proof Let $B = \{b_{ij}\}$ be fa-consistent. Then, by setting $a_{ij} = \varphi_\sigma^{-1}(b_{ij})$, we obtain

$$a_{ij}a_{jk}a_{ki} = \sigma^{2b_{ij}-1}\sigma^{2b_{jk}-1}\sigma^{2b_{ki}-1} = \sigma^{2(b_{ij}+b_{jk}+b_{ki})-3} = \sigma^{2(\frac{3}{2})-3} = 1$$

for all $i, j, k \in \{1, 2, ..., n\}$. Therefore, $A = \{a_{ij}\}$ is m-consistent.   □

Now, let us define the function ϕ and its inverse ϕ^{-1} as follows:

$$\phi(t) = \frac{t}{1+t} \text{ for } t > 0 , \tag{2.87}$$

$$\phi^{-1}(t) = \frac{t}{1-t} \text{ for } 0 < t < 1 . \tag{2.88}$$

We obtain the following results, characterizing fm-consistent and m-consistent matrices, see [81].

Proposition 2.23 *Let* $A = \{a_{ij}\}$ *be an* $n \times n$ *matrix with* $0 < a_{ij}$ *for all* i *and* j.
If $A = \{a_{ij}\}$ *is m-consistent then* $B = \{b_{ij}\} = \{\phi(a_{ij})\}$ *is fm-consistent.*

Proof By (2.87) and m-reciprocity (2.1) we obtain for all i, j:

$$b_{ij} = \frac{a_{ij}}{1 + a_{ij}} = \frac{\frac{1}{a_{ji}}}{1 + \frac{1}{a_{ji}}} = \frac{1}{1 + a_{ji}} = \frac{a_{ji}}{1 + a_{ji}} a_{ij} = b_{ji} a_{ij} .$$

Then, $b_{ij} b_{jk} b_{ki} = b_{ji} b_{kj} b_{ik} a_{ij} a_{jk} a_{ki}$, for all i, j and k. Hence, by (2.3), B is fm-consistent. $\qquad\square$

Proposition 2.24 *Let* $B = \{b_{ij}\}$ *be an f-reciprocal* $n \times n$ *matrix with* $0 < b_{ij} < 1$
for all i *and* j.
If $B = \{b_{ij}\}$ *is fm-consistent then* $A = \{a_{ij}\} = \{\phi^{-1}(b_{ij})\}$ *is m-consistent.*

Proof Let B be f-reciprocal. Then

$$a_{ij} = \frac{b_{ij}}{1 - b_{ij}} = \frac{b_{ij}}{b_{ji}} ,$$

Then by fm-consistency, we obtain

$$a_{ij} a_{jk} a_{ki} = \frac{b_{ij}}{b_{ji}} \frac{b_{jk}}{b_{kj}} \frac{b_{ki}}{b_{ik}} = 1.$$

Hence, $A = \{a_{ij}\}$ is m-consistent. $\qquad\square$

From Proposition 2.22 it is clear that the concept of fm-consistency plays a similar role for f-reciprocal PCF matrices as the concept of m-consistency does for m-reciprocal matrices.

2.9.2 Methods for Deriving Priorities from PCF Matrices

A parallel result to Proposition 2.1 can be derived for f-reciprocal matrices.

Proposition 2.25 *Let* $A = \{a_{ij}\}$ *be an f-reciprocal* $n \times n$ *matrix with* $0 < a_{ij} < 1$
for all i *and* j.
$A = \{a_{ij}\}$ is fm-consistent if and only if there exists a vector $v = (v_1, v_2, ..., v_n)$
with $v_i > 0$ *for all* $i = 1, 2, ..., n$, *and*

$$\sum_{j=1}^{n} v_j = 1$$

such that

$$a_{ij} = \frac{v_i}{v_i + v_j} \text{ for all } i, j \in \{1, \cdots, n\} . \tag{2.89}$$

Proof By Propositions 2.23 and 2.24, $B = \{b_{ij}\}$ is fm-consistent if and only if $A = \{a_{ij}\} = \{\frac{b_{ij}}{1-b_{ij}}\}$ is m-consistent. By Proposition 2.3, this result is equivalent to the existence of a vector $v = (v_1, v_2, \cdots, v_n)$ with $v_i > 0$ for all $i \in \{1, \cdots, n\}$, such that $\frac{b_{ij}}{1-b_{ij}} = \frac{v_i}{v_j}$ for all $i, j \in \{1, \cdots, n\}$, or, equivalently, $b_{ij} = \frac{v_i}{v_i+v_j}$ for all $i, j \in \{1, \cdots, n\}$, i.e. (2.89) is true.   □

Here, the vector $v = (v_1, v_2, \cdots, v_n)$ with $v_i > 0$ for all $i = 1, 2, ..., n$, and $\sum_{j=1}^{n} v_j = 1$ is called the *fm-priority vector of A*.

The previous result is a representation theorem for fm-consistent matrices, the next two results concern fa-consistent matrices.

Proposition 2.26 *Let $A = \{a_{ij}\}$ be an $n \times n$ matrix with $0 < a_{ij} < 1$ for all $i, j \in \{1, 2, ..., n\}$. $A = \{a_{ij}\}$ is fa-consistent if and only if*

$$a_{ij} = \frac{1}{2}(1 + nu_i - nu_j) \text{ for all } i, j \in \{1 \cdots, n\} , \tag{2.90}$$

where

$$u_i = \frac{2}{n^2} \sum_{k=1}^{n} a_{ik} \text{ for all } i \in \{1, \cdots, n\} , \tag{2.91}$$

and

$$\sum_{i=1}^{n} u_i = 1 . \tag{2.92}$$

Proof (i) Let let $A = \{a_{ij}\}$ be fa-consistent, then A is f-reciprocal and by (2.79) we obtain

$$\sum_{i,j=1}^{n} a_{ij} = \frac{1}{2}n^2 . \tag{2.93}$$

From (2.91), (2.93) we obtain

$$\sum_{i=1}^{n} u_i = \frac{2}{n^2} \sum_{i,j=1}^{n} a_{ij} = 1 . \tag{2.94}$$

From (2.91), (2.94) we also have

$$\frac{1}{2} + \frac{n}{2}(u_i - u_j) = \frac{1}{2} + \frac{1}{n}\sum_{k=1}^{n}(a_{ik} - a_{jk}) \text{ for all } i, j \in \{1, \cdots, n\} .$$

By fa-consistency of B, (2.82), we obtain

$$a_{ij} - \frac{1}{2} = a_{ik} - a_{jk} \text{ for all } i, j \in \{1, \cdots, n\} .$$

If (2.7) is satisfied for all $i, j \in \{1, \cdots, n\}$, then

$$a_{ij} = \frac{1}{2}(1 + nu_i - nu_j) \text{ for all } i, j \in \{1, \cdots, n\} .$$

(ii) Let (2.90) be true, then easily (2.81) is easily satisfied. $\qquad\square$

Here, the vector $u = (u_1, u_2, \cdots, u_n)$ with $u_i > 0$ for all $i = 1, \cdots, n$, and $\sum_{j=1}^{n} u_j = 1$ is called the *fa-priority vector* of A.

By the symbol I_{mY} we denote the m-consistency index that has been defined for various methods (eigenvalue, least squares, geometric mean) for deriving priorities of multiplicative pairwise comparison matrices, where $Y \in \{EV, LS, GM\}$, i.e., I_{mEV}, I_{mLS}, I_{mGM}.

Now, we shall investigate inconsistency property also for f-reciprocal matrices. For this purpose, we use relations between m-consistent and fm-consistent/fa-consistent matrices derived in Propositions 2.21–2.24.

Let $B = \{b_{ij}\}$ be an f-reciprocal $n \times n$ matrix with $0 < b_{ij} < 1$ for all $i, j \in \{1, \cdots, n\}$, and let $Y \in \{EV, LS, GM\}$. We define the *faY-consistency index* $I_{faY}(B)$ of $B = \{b_{ij}\}$ as

$$I_{faY}(B) = \phi(I_{mY}(A)), \text{ where } A = \{\phi^{-1}(b_{ij})\} . \qquad (2.95)$$

From (2.95) we obtain the following result which is parallel to Proposition 2.3.

Proposition 2.27 *If $B = \{b_{ij}\}$ is an f-reciprocal $n \times n$ PCF matrix with $0 < b_{ij} < 1$ for all $i, j = 1, \cdots, n$, $Y \in \{EV, LS, MG\}$. Then $I_{faY}(B) \geq 0$. Moreover, B is fa-consistent if and only if $I_{faY}(B) = 0$.*

Proof The proof follows from the corresponding proofs of the propositions in Sect. 2.4. $\qquad\square$

Further, we shall be dealing with the inconsistency of fm FPC matrices. Recall transformation functions φ_σ and φ_σ^{-1} defined by (2.85), (2.86), where $\sigma > 1$ is a given scale value. Let $B = \{b_{ij}\}$ be an f-reciprocal $n \times n$ matrix with $0 < b_{ij} < 1$ for all $i, j = 1, \cdots, n$, $Y \in \{EV, LS, MG\}$. We define the *fmY-consistency index* $I_{fmY}^{\sigma}(B)$ of $B = \{b_{ij}\}$

$$I_{fmY}^{\sigma}(B) = \varphi_\sigma(I_{mY}(A_\sigma)), \text{ where } A_\sigma = \{\varphi_\sigma^{-1}(b_{ij})\} . \qquad (2.96)$$

From (2.86), (2.96) we obtain the following result which is parallel to Propositions 2.3 and 2.27.

Proposition 2.28 *If $B = \{b_{ij}\}$ is an f-reciprocal $n \times n$ matrix with $0 < b_{ij} < 1$ for all $i, j \in \{1, \cdots, n\}$, $Y \in \{EV, LS, MG\}$, then $I^{\sigma}_{fmY}(B) \geq 0$. Moreover, B is fm-consistent if and only if $I^{\sigma}_{fmY}(B) = 0$.*

Proof The proof follows from the corresponding proofs of the propositions in Sect. 2.4.   □

Example 2.8 Let $\mathscr{C} = \{c_1, c_2, c_3, c_4\}$ be the set of 4 alternatives. The preferences on $\mathscr{C}$ are described by the PCF matrix $B = \{b_{ij}\}$, where

$$B = \begin{pmatrix} 0.5\ 0.6\ 0.6\ 0.9 \\ 0.4\ 0.5\ 0.6\ 0.7 \\ 0.4\ 0.4\ 0.5\ 0.5 \\ 0.1\ 0.3\ 0.5\ 0.5 \end{pmatrix}. \tag{2.97}$$

Here, $B = \{b_{ij}\}$ is f-reciprocal and fm-inconsistent, as it may be directly verified by (2.84), e.g., $b_{12}.b_{23}.b_{31} \neq b_{32}.b_{21}.b_{13}$. At the same time, B is fa-inconsistent as $b_{12} + b_{23} + b_{31} = 1.9 \neq 1.5$. Now, we consider the scale $[1/\sigma, \sigma]$ with $\sigma = 9$. Then we calculate

$$E = \{\phi^{-1}(b_{ij})\} = \begin{pmatrix} 1 & 1.50\ 1.50\ 9.00 \\ 0.67 & 1 & 1.5 & 2.33 \\ 0.67\ 0.67 & 1 & 1 \\ 0.11\ 0.43 & 1 & 1 \end{pmatrix},$$

$$F = \{\varphi_9^{-1}(b_{ij})\} = \begin{pmatrix} 1 & 1.55\ 1.55\ 5.80 \\ 0.64 & 1 & 1.55 & 2.41 \\ 0.64\ 0.64 & 1 & 1 \\ 0.17\ 0.42 & 1 & 1 \end{pmatrix}.$$

Further, we calculate the maximal eigenvalues $\rho(E) = 4.29$ and $\rho(F) = 4.15$.

By (2.12), (2.96) we obtain $I_{fmEV}(B) = 0.11$ with the priority vector: $w^{fmEV} = (0.47, 0.25, 0.18, 0.10)$, which gives the ranking of the alternatives: $c_1 \succ c_2 \succ c_3 \succ c_4$.

Similarly, $I^9_{faEV}(B) = 0.056$ with the priority vector: $w^{faEV} = (0.44, 0.27, 0.18, 0.12)$, giving the same ranking of alternatives: $c_1 \succ c_2 \succ c_3 \succ c_4$.

2.10 Conclusion

In this chapter, pairwise comparison is understood as any process of comparing alternatives in pairs to judge which of each alternative is preferred, or has a greater amount of some quantitative property, or whether or not the two alternatives are identical, or indifferent. The entry of the preference matrix can assume different meanings: it can be a preference ratio (multiplicative case) or a preference difference (additive case), or it belongs to the unit interval [0; 1] and measures the distance from the indifference that is expressed by 0.5 (fuzzy case). When comparing two elements, the decision-maker assigns the value from a scale to any pair of alternatives representing the element of the pairwise preference matrix. Here, we investigated particularly the relations between the transitivity and consistency of preference matrices being understood differently with respect to the type of preference matrix. By various methods for deriving priorities from various types of preference matrices we obtained the corresponding priority vectors for final ranking of alternatives. Finally, various properties of priority vectors determining the ranking of the original alternatives were investigated. Illustrative numerical examples were provided and discussed.

References

1. Aczel J (1966) Lectures on functional equation and their applications. Academic Press, New York
2. Aguaron J, Moreno-Jimenez JM (2003) The geometric consistency index: approximated thresholds. Eur J Oper Res 147(1):137–145
3. Alonso JA, Lamata MT (2006) Consistency in the analytic hierarchy process: a new approach. Int J Uncertain Fuzziness Knowledge-Based Syst 14(4):445–459
4. Alonso S, Chiclana F, Herrera F, Herrera-Viedma E, Alcala-Fdes J, Porcel C (2008) A consistency-based procedure to estimate missing pairwise preference values. Int J Intell Syst 23:155–175
5. Bana e Costa CA, Vansnick J (2008) A critical analysis of the eigenvalue method udes to derive priorities in AHP. Eur J Oper Res 187(3):1422–1428
6. Barzilai J (1997) Deriving weights from pairwise comparison matrices. J Oper Res Soc 48(12):1226–1232
7. Barzilai J (1998) Consistency measures for pairwise comparison matrices. J Multicrit Decis Anal 7:123–132
8. Barzilai J (2001) Notes on the analytic hierarchy process. In: Proceedings of the NSF design and manufacturing research conference, Tampa, Florida, pp 1–6
9. Barzilai J (2003) Preference function modeling: the mathematical foundations of decision theory. In: Ehrgott M et al Trends in multiple criteria decision analysis. Springer, Berlin, pp 57–86
10. Blankmeyer E (1987) Approaches to consistency adjustments. J Optim Theory Appl 154:479–488
11. Bourbaki N (1990) Algebra II. Springer, Berlin
12. Bollobas J (2002) Modern graph theory. Springer, Berlin
13. Boyd S, Vandenberghe L (2004) Convex optimization. Cambridge University Press, Cambridge
14. Brin S, Page L (1998) The anatomy of a large-scale hypertextual web search engine. Comput Netwo ISDN Syst 30:107–117
15. Brunelli M, Canal L, Fedrizzi M (2013) Inconsistency indices for pairwise comparison matrices: a numerical study. Ann Oper Res 211(1):493–509
16. Brunelli M, Fedrizzi M (2015) Axiomatic properties of inconsistency indices for pairwise comparisons. J Oper Res Soc 66(1):1–15
17. Brunelli M (2017) Studying a set of properties of inconsistency indices for pairwise comparisons. Ann Oper Res 248(1,2):143–161

18. Cavallo B, DApuzzo L (2009) A general unified framework for pairwise comparison matrices in multicriterial methods. Int J Intell Syst 24(4):377–398

19. Cavallo B, DApuzzo L (2010) Characterizations of consistent pairwise comparison matrices over abelian linearly ordered groups. Int J Intell Syst 25:1035–1059

20. Cavallo B, DApuzzo L, Squillante M (2012) About a consistency index for pairwise comparison matrices over a divisible alo-group. Int J Intell Syst 27:153–175

21. Cavallo B, DApuzzo L (2012) Deriving weights from a pairwise comparison matrix over an alo/group. Soft Comput 16:353–366

22. Cavallo B, D'Apuzzo L (2015) Reciprocal transitive matrices over Abelian linearly ordered groups: characterizations and application to multicriteria decision problems. Fuzzy Sets Syst 266:33–46

23. Cavallo B, D'Apuzzo L (2016) Ensuring reliability of the weighting vector: weak consistent pairwise comparison matrices. Fuzzy Sets Syst 296:21–34

24. Colomer JM (2011) Ramon Llull: from Ars electionis to social choice theory. Soc Choice Welf 40(2):317–328

25. Condorcet M (1785) Essay on the application of analysis to the probability of majority decisions. Imprimerie Royale, Paris

26. Grabisch M, Marichal JL, Mesiar R, Pap L (2009) Aggregation functions. Cambridge University Press, New York

27. Chang JSK et al (2008) Note on deriving weights from pairwise comparison matrices in AHP. Inf Manag Sci 19(3):507–517

28. Chen Q, Triantaphillou E (2001) Estimating data for multicriteria decision making problems: optimization techniques. In: Pardalos PM, Floudas C (eds) Encyclopedia of optimization, vol. 2. Kluwer Academic Publishers, Boston

29. Chiclana F, Herrera F, Herrera-Viedma E (2001) Integrating multiplicative preference relations in a multipurpose decision making model based on fuzzy preference relations. Fuzzy Sets Syst 112:277–291

30. Chiclana F, Herrera F, Herrera-Viedma E, Alonso S (2007) Some induced ordered weighted averaging operators and their use for solving group decision-making problems based on fuzzy preference relations. Eur J Oper Res 182:383–399

31. Chiclana F, Herrera-Viedma E, Alonso S (2009) A note on two methods for estimating pairwise preference values. IEE Trans Syst Man Cybern 39(6):1628–1633

32. Choo E, Wedley W (2004) A common framework for deriving preference values from pairwise comparison matrices. Comput Oper Res 31(6):893–908

33. Cook W, Kress M (1988) Deriving weights from pairwise comparison ratio matrices: an axiomatic approach. Eur J Oper Res 37(3):355–362

34. Crawford G, Williams C (1985) A note on the analysis of subjective judgment matrices. J Math Psychol 29:25–40

35. Dopazo E, Gonzales-Pachon J (2003) Consistency-driven approximation of a pairwise comparison matrix. Kybernetika 39(5):561–568

36. Dubois D, Prade H (1980) Fuzzy sets and systems: theory and application. Academic, New York

37. Fan ZP, Ma J, Zhang Q (2002) An approach to multiple attribute decision making based on fuzzy preference information on alternatives. Fuzzy Sets Syst 131:101–106

38. Fan ZP, Hu GF, Xiao SH (2004) A method for multiple attribute decision-making with the fuzzy preference relation on alternatives. Comput Ind Eng 46:321–327

39. Fan ZP, Ma J, Jiang YP, Sun YH, Ma L (2006) A goal programming approach to group decision making based on multiplicative preference relations and fuzzy preference relations. Eur J Oper Res 174:311–321

40. Fedrizzi M, Giove S (2007) Incomplete pairwise comparison and consistency optimization. Eur J Oper Res 183(1):303–313

41. Fedrizzi M, Brunelli M (2010) On the priority vector associated with a reciprocal relation and a pairwise comparison matrix. Soft Comput 14(2):639–645

42. Fiedler M, Nedoma J, Ramík J, Rohn J, Zimmermann K (2006) Linear optimization problems with inexact data. Springer, Berlin
43. Figueira J, Greco S, Ehrgott M (2005) Multiple criteria decision analysis: state of the art surveys. Springer, New York
44. Fodor J, Roubens M (1994) Fuzzy preference modeling and multicriteria decision support. Kluwer, Dordrecht
45. Gantmacher FR (1959) The theory of matrices, vol 1. Chelsea
46. Gavalec M, Ramik J, Zimmermann K (2014) Decision making and optimization - special matrices and their applications in economics and management. Springer International Publishing, Cham
47. Gong ZW (2008) Least-square method to priority of the fuzzy preference relations with incomplete information. Int J Approx Reason 47:258–264
48. Gong Z, Li L, Cao J, Zhou F (2010) On additive consistent properties of the intuitionistic fuzzy preference relation. Int J Inform Technol Decis Mak 9(6):1009–1025
49. Harker PT (1987) Alternative modes of questioning in the analytic hierarchy process. Math Modell 9(35):353–360
50. Harker PT (1987) Incomplete pairwise comparisons in the analytic hierarcy process. Math Modell 9(11):837–848
51. Herrera-Viedma E, Herrera F, Chiclana F, Luque M (2004) Some issues on consistency of fuzzy preference relations. Eur J Oper Res 154:98–109
52. Herrera F, Herrera-Viedma E, Chiclana F (2001) Multiperson decision making based on multiplicative preference relations. Eur J Oper Res 129:372–385
53. Herrera F, Herrera-Viedma E (2002) Choice functions and mechanisms for linguistic preference relations. Eur J Oper Res 120:144–161
54. Herrera-Viedma E, Chiclana F, Herrera F, Alonso S (2007) A group decision-making model with incomplete fuzzy preference relations based on additive consistency. IEEE Trans Syst Man Cybern Part B 37:176–189
55. Herrera-Viedma E, Alonso S, Chiclana F, Herrera F (2007) A consensus model for group decision-making with incomplete fuzzy preference relations. IEEE Trans Fuzzy Syst 15(5):863–877
56. Hovanov NV, Kolari JW, Sokolov MV (2008) Deriving weights from general pairwise comparison matrices. Math Soc Sci 55:205–220
57. International MCDM Society. http://www.mcdmsociety.org/
58. Koczkodaj WW (1993) A new definition of consistency of pairwise comparisons. Math Comput Model 18(7):79–84
59. Koczkodaj WW, Magnot JP, Mazurek J, Peters JF, Rakhshani H, Soltys M, Strzalka D, Szybowski J, Tozzi A (2017) On normalization of inconsistency indicators in pairwise comparisons. Int J Approx Reason 86:73–79
60. Kou G, Ergu D, Lin CS, Chen Y (2016) Pairwise comparison matrix in multiple criteria decision making. Technol Econ Dev Econ 22(5):738–765
61. Kulakowski K (2015) Notes on order preservation and consistency in AHP. Eur J Oper Res 245:333–337
62. Kulakowski K, Mazurek J, Ramik J, Soltys M (2019) When is the condition of preservation met? Eur J Oper Res 277(1):248–254
63. Kim SH, Choi SH, Kim JK (1999) An interactive procedure for multiple attribute group decision making with incomplete information: Range-based approach. Eur J Oper Res 118:139–152
64. Lee HS (2006) A fuzzy method for evaluating suppliers. In: Wang L et al (eds) FSKD, LNAI 4223. Springer, Berlin, pp 1035–1043
65. Lee HS, Tseng WK (2006) Goal programming methods for constructing additive consistency fuzzy preference relations. In: Gabrys B et al (eds) KES 2006, Part II, LNAI 4252. Springer, Berlin, pp 910–916
66. Lee HS, Yeh CH (2008) Fuzzy multi-criteria decision making based on fuzzy preference relation. In: Lovrek I et al (eds) KES, Part II, LNAI 5178. Springer, Berlin, pp 980–985

67. Lee HS, Shen PD, Chyr WL (2008) Prioritization of incomplete fuzzy preference relation. In: Lovrek I et al (eds) KES, Part II, LNAI 5178. Springer, Berlin, pp 974–979

68. Lin CC (2007) A revised framework for for deriving preference values from pairwise comparison matrices. Eur J Oper Res 176:1145–1150

69. Lipovetsky S, Conklin MW (2002) Robust estimation of priorities in AHP. Eur J Oper Res 137:110–122

70. Lundy M, Siraj S, Greco S (2017) The mathematical equivalence of the 'spanning tree' and row geometric mean preference vectors and its implications for preference analysis. Eur J Oper Res 257:197–208

71. Ma J et al (2006) A method for repairing the inconsistency of fuzzy preference relations. Fuzzy Sets Syst 157:20–33

72. Mazurek J, Perzina R (2017) On the inconsistency of pairwise comparisons: an experimental study. Scientific papers of the University of Pardubice. Ser D3 Fac Econ Adm 41:102–109

73. Mazurek J, Ramik J (2019) Some new properties of inconsistent pairwise comparisons matrices. Int J Approx Reason 113:119–132

74. Mikhailov L (2000) A fuzzy programming method for deriving priorities in the analytic hierarchy process. J Oper Res Soc 5:342–349

75. Nishizawa K (2004) Estimation of unknown comparisons in incomplete AHP and its compensation. Report of the Research Institute of Industrial Technology, Nihon University, 77

76. Nocedal J, Wright SJ (2006) Numerical optimization, 2nd edn. Springer, Berlin

77. Pelaez JI, Lamata MT (2003) A new measure of inconsistency for positive reciprocal matrices. Comput Math Appl 46(12):1839–1845

78. Ramík J, Vlach M (2001) Generalized concavity in optimization and decision making. Kluwer Publishing Company, Boston

79. Ramík J, Korviny P (2010) Inconsistency of pairwise comparison matrix with fuzzy elements based on geometric mean. Fuzzy Sets Syst 161:1604–1613

80. Ramík J, Vlach M (2012) Aggregation functions and generalized convexity in fuzzy optimization and decision making. Ann Oper Res 191:261–276

81. Ramík J, Vlach M (2013) Measuring consistency and inconsistency of pair comparison systems. Kybernetika 49(3):465–486

82. Ramík J (2014) Incomplete fuzzy preference matrix and its application to ranking of alternatives. Int J Intell Syst 29(8):787–806

83. Ramik J (2015) Isomorphisms between fuzzy pairwise comparison matrices. Fuzzy Optim Decis Mak 14:199–209

84. Ramik J (2015) Pairwise comparison matrix with fuzzy elements on alo-groups. Inf Sci 297:236–253

85. Ramik J (2017) Ranking alternatives by pairwise comparisons matrix and priority vector. Sci Ann Econ Bus 64(SI):85–95

86. Ramik J (2018) Strong reciprocity and strong consistency in pairwise comparison matrix with fuzzy elements. Fuzzy Optim Decis Mak 17:337–355

87. Lull R Artifitium electionis personarum (The method for the elections of persons), pp 1274–1283. https://www.math.uni-augsburg.de/htdocs/emeriti/pukelsheim/llull/

88. Roy B (1968) Classement et choix en prsence de points de vue multiples (la mthode ELECTRE). RIRO 8:5775

89. Roy B (1990) Decision-aid and decision-making. Eur J Oper Res 45:324–331

90. Rubinov A (2000) Abstract convexity and global optimization. Springer Science + Business Media B.V, Boston

91. Saaty TL (1977) A scaling method for priorities in hierarchical structures. J Math Psychol 15(3):234–281

92. Saaty TL (1980) The analytic hierarchy process. McGraw-Hill, New York

93. Saaty TL (1994) Fundamentals of decision making and priority theory with the AHP. RWS Publications, Pittsburgh

94. Saaty TL, Vargas L (2000) Models, methods, concepts and applications of the analytic hierarchy process. Kluwer, Boston

95. Saaty L (2003) Decision-making with the AHP: why is the principal eigenvector necessary. Eur J Oper Res 145:85–91
96. Saaty TL (2004) Decision making the analytic hierarchy and network processes (AHP/ANP). J Syst Sci Syst Eng 13(1):1–34
97. Saaty TL (2008) Decision making with the analytic hierarchy process. Int J Serv Sci 1:83–98
98. Shiraishi S, Obata T, Daigo M (1998) Properties of a positive reciprocal matrix and their application to AHP. J Oper Res Soc Jpn 41(3):404–414
99. Srdjevic B (2005) Combining different prioritization methods in the analytic hierarchy process synthesis. Comput Oper Res 32:1897–1919
100. Switalski Z (2003) General transitivity conditions for fuzzy reciprocal preference matrices. Fuzzy Sets Syst 137:85–100
101. Tanino T (1984) Fuzzy preference orderings in group decision making. Fuzzy Sets Syst 12:117–131
102. Tanino T (1988) Fuzzy preference relations in group decision making. In: Kacprzyk J, Roubens M (eds) Non-conventional preference relations in decision making. Springer, Berlin, pp 54–71
103. Thurstone LL (1927) A law of comparative judgment. Psychol Rev 34:273–286
104. Xu ZS (2004) On compatibility of interval fuzzy preference relations. Fuzzy Optim Decis Mak 3:217–225
105. Xu ZS (2004) Goal programming models for obtaining the priority vector of incomplete fuzzy preference relation. Int J Approx Reason 36:261–270
106. Xu ZS, Da QL (2005) A least deviation method to obtain a priority vector of a fuzzy preference relation. Eur J Oper Res 164:206–216
107. Xu ZS, Chen J (2008) Some models for deriving the priority weights from interval fuzzy preference relations. Eur J Oper Res 184:266–280
108. Xu ZS (2005) A procedure for decision making based on incomplete fuzzy preference relation. Fuzzy Optim Decis Mak 4:175–189
109. Xu ZS (2007) Multiple-attribute group decision making with different formats of preference information on attributes. IEEE Trans Syst Man Cybern B 37:1500–1511
110. Xu ZS, Chen J (2007) A subjective and objective integrated method for MAGDM problems with multiple types of exact preference formats. In: IDEAL, pp 145–154
111. Xu ZS, Chen J (2008) Group decision-making procedure based on incomplete reciprocal relations. Soft Comput 12:515–521
112. Xu ZS, Chen J (2008) MAGDM linear-programming models with distinct uncertain preference structures. IEEE Trans Syst Man Cybern B 38:1356–1370
113. Yuen KKF (2012) Pairwise opposite matrix and its cognitive prioritization operators: comparisons with pairwise reciprocal matrix and analytic prioritization operators. J Oper Res Soc 63:322–338
114. Wang YM, Fan ZP (2007) Group decision analysis based on fuzzy preference relations: logarithmic and geometric least squares methods. Appl Math Comput 194:108–119
115. Wang YM, Parkan C (2005) Multiple attribute decision making based on fuzzy preference information on alternatives: ranking and weighting. Fuzzy Sets Syst 153:331–346
116. Wang YM, Fan ZP, Hua Z (2007) A chi-square method for obtaining a priority vector from multiplicative and fuzzy preference relations. Eur J Oper Res 182:356–366

Chapter 3
Pairwise Comparisons Matrices on Alo-Groups in Decision-Making

3.1 Unified Framework for Pairwise Comparisons Matrices over ALO-Groups

3.1.1 Introduction

As mentioned before, a pairwise comparison matrix $A = \{a_{ij}\}$ is a helpful tool to determine the ranking on a set $\mathscr{C} = \{c_1, c_2, \cdots, c_n\}$ of alternatives (criteria, objects, or any other entities). The entry a_{ij} of the matrix $A = \{a_{ij}\}$ assumes three different meanings:

- a_{ij} can be a preference ratio (multiplicative case), or
- a_{ij} can be a preference difference (additive case), or
- a_{ij} belongs to the unit interval $[0; 1]$ and measures the distance from the indifference that is expressed by 0.5 (fuzzy case).

In this section we consider pairwise comparison matrices over an *abelian linearly (totally) ordered group* and, in this way, we provide a general framework for all the abovementioned cases. By introducing a more general setting, we provide an optimization method for generating priority vector satisfying appropriate conditions. Then we deal with the problem of measuring the inconsistency of the PCM that has a natural meaning, and it corresponds to the consistency indexes presented in the previous chapter. In some sense we unify the major part of the theory presented in Chap. 2. Moreover, we investigate some possibilities for weakening the consistency condition of the PCM introducing weak consistency, transitivity, and strong transitivity. Finally, we deal with PCMs with some missing elements. The matter of this chapter is based on the textbook by Bourbaki, [5], and simple and elegant ideas in the papers of Cavallo and D'Apuzzo [8–10]. Some elements of abelian linearly (totally) ordered groups are summarized in Chap. 1.

3.1.2 Continuous Alo-Groups over a Real Interval

Let us briefly recall some useful basic concepts from the theory of groups (see also Chap. 1).

An *abelian group* is a set, G, together with an operation $\odot$ (read: "odot") that combines any two elements $a, b \in G$ to form another element in G denoted by $a \odot b$, see [5, 9]. The symbol $\odot$ is a general placeholder for some concretely given operation. $(G, \odot)$ satisfies the following requirements known as the *abelian group axioms*, particularly: *commutativity*, and *associativity*, there exists an *identity element* $e \in G$ and for each element $a \in G$ there exists an element $a^{(-1)} \in G$ called the *inverse element to a*. The *inverse operation* $\div$ to $\odot$ is defined for all $a, b \in G$ as follows: $a \div b = a \odot b^{(-1)}$. Note that the inverse operation is not necessarily associative.

An ordered triple $(G, \odot, \le)$ is said to be an *abelian linearly ordered group, alo-group* for short, if $(G, \odot)$ is a group, $\le$ is a linear order on G, and for all $a, b, c \in G$: $a \le b$ implies $a \odot c \le b \odot c$. In other words, it respects $\le$.

If $\mathscr{G} = (G, \odot, \le)$ is an alo-group, then G is naturally equipped with the order topology induced by $\le$ and $G \times G$ is equipped with the related product topology. We say that $\mathscr{G}$ is a *continuous alo-group* if the operation $\odot$ is continuous on $G \times G$.

By definition, an alo-group $\mathscr{G}$ is a lattice-ordered group. Hence, there exists $\max\{a, b\}$, for each pair $(a, b) \in G \times G$. Nevertheless, a nontrivial alo-group $\mathscr{G} = (G, \odot, \le)$ has neither the greatest element nor the least element. Because of the associative property, the operation $\odot$ can be extended by induction to an *n-ary* operation.

$\mathscr{G} = (G, \odot, \le)$ is *divisible* if for each positive integer n and each $a \in G$ there exists the *(n)th root* of a denoted by $a^{(1/n)}$, i.e., $\left(a^{(1/n)}\right)^{(n)} = a$.

The function $\|.\| : G \to G$ defined for each $a \in G$ by

$$\|a\| = \max\{a, a^{(-1)}\} \tag{3.1}$$

is called a *$\mathscr{G}$-norm*. In the sequel we shall assume that G is a divisible and continuous alo-group. Then G is an open interval in $\mathbf{R}$.

The operation $d : G \times G \to G$ defined by $d(a, b) = \|a \div b\|$ for all $a, b \in G$ is called a *$\mathscr{G}$-distance*. Next, we present the well-known examples of alo-groups; for more details see also [9], or [32].

Let G be a subset of the real line $\mathbf{R} =] - \infty, +\infty[$, $\le$ be the total order on G inherited from the usual order on $\mathbf{R}$ and $\mathscr{G} = (G, \odot, \le)$ be a real alo-group. Let G be a proper interval of $\mathbf{R}$. By (1.21) and (1.22), G is an open interval.

By $\mathbf{Q}$ we denote the set of the rational numbers, $\mathbf{Q}^+$ is the set of all positive rational numbers, operation $+$ is the usual addition and $\cdot$ is the usual multiplication on $\mathbf{R}$. We obtain the following examples of real alo-groups (see [9, 10]).

Example 3.1 Let $\mathscr{R} = (\mathbf{R}, +, \le)$ and $\mathscr{Q} = (\mathbf{Q}, +, \le)$ be two alo-groups. Both alo-groups called *additive alo-groups* are continuous with: $e = 0$, $a^{(-1)} = -a$, $a^{(n)} = na$, $a \div b = a - b$; the norm $\|a\| = |a| = a \vee (-a)$ generates the usual distance over $\mathbf{R}$ (resp. $\mathbf{Q}$):

$$|a - b| = (a - b) \vee (b - a).$$

Both **R** and **Q** are divisible, and the (n)-root of $x^{(n)} = a$ is the solution of $nx = a$, usually indicated by the symbol a/n. The mean $m_\odot(a_1, a_2, \cdots, a_n)$ is the usual arithmetic average, i.e., $m_\odot(a_1, a_2, \cdots, a_n) = \frac{1}{n} \sum a_i$.

Example 3.2 $\mathscr{R}_+ = (]0, +\infty[, \cdot, \leq)$ and $\mathscr{Q}^+ = (\mathbf{Q}^+, \cdot, \leq)$ are two continuous alo-groups with: $e = 1, x^{(-1)} = x^{-1} = 1/x$, $x^{(n)} = x^n$, $x \div y = \frac{x}{y}$, $\|a\| = a \vee a^{-1}$; and both $d_{\mathscr{R}_+}(a, b)$ and $d_{\mathscr{Q}^+}(a, b)$ are given by $\|a \div b\| = \|\frac{a}{b}\| = \frac{a}{b} \vee \frac{b}{a} \in [1, +\infty[$. The alo-group $\mathscr{R}_+$ called *multiplicative alo-group* is divisible and the (n)-root of a is $x = a^{1/n}$. The mean $m.(a_1, a_2, \cdots, a_n)$ is the geometric mean, i.e.,

$$m.(a_1, a_2, \cdots, a_n) = \left(\prod_{i=1}^{n} a_i \right)^{1/n}.$$

The alo-group $\mathscr{Q}^+$ is not divisible as, for example, $x^2 = 2$ has no solution in $\mathbf{Q}^+$.

Example 3.3 $]0; 1[_m = (]0; 1[, \cdot_f, \leq)$ is *fuzzy-multiplicative alo-group* with

$$a \cdot_f b = \frac{a \cdot b}{a \cdot b + (1 - a) \cdot (1 - b)}, \quad e = 0, 5, \ a^{(-1)} = 1 - a,$$

$$\|a\| = max\{a, 1 - a\}.$$

Here,

$$\|a \div_f b\| = max\{\frac{a \cdot (1 - b)}{a \cdot (1 - b) + (1 - a) \cdot b}, \frac{(1 - a) \cdot b}{a \cdot (1 - b) + (1 - a) \cdot b}\} \in]0; 1[.$$
$$(3.2)$$

We obtain also the mean and product

$$m_f(a_i; i \in \{1, \cdots, n\}) = \frac{(\prod_{i=1}^{n} a_i)^{1/n}}{(\prod_{i=1}^{n} a_i)^{1/n} + (\prod_{i=1}^{n}(1 - a_i))^{1/n}}. \qquad (3.3)$$

$$a_1 \cdot_f \ldots \cdot_f a_n = \frac{\prod_{i=1}^{n} a_i}{\prod_{i=1}^{n} a_i + \prod_{i=1}^{n}(1 - a_i)}. \qquad (3.4)$$

Particularly,

$$a^{(n)} \frac{a^n}{a^n + (1 - a)^n}. \qquad (3.5)$$

The fuzzy-multiplicative alo-group $]0; 1[_m$ is divisible and continuous. For more details and properties, see [10].

Example 3.4 Fuzzy-additive alo-group $\mathscr{R}_a=(] - \infty; +\infty[, +_f, \leq)$ is a continuous alo-group with

$$a +_f b = a + b - 0.5, \ e = 0.5, \ a^{(-1)} = 1 - a,$$

$$a^{(n)} = n.a - \frac{n-1}{2}, \ a -_f b = a - b + 0.5, \ \|a\| = max\{a, 1 - a\} \ ,$$

and

$$d_{]0;1[_a}(a, b) = \|a -_f b\| = max\{a - b + 0.5, b - a - 0.5\} \ .$$

The fuzzy-additive alo-group $\mathscr{R}_a$ is divisible and the (n)-root of a is $a^{(\frac{1}{n})} = a/n$.

The mean $m_{+_f}(a_1, a_2, \cdots, a_n) = \frac{1}{n} \sum_{i=1}^{n} a_i$. The entries of a PC matrix $A = \{a_{ij}\}$ are taken from the unit interval $[0; 1]$ with a fuzzy interpretation. A is *fa-consistent* if

$$a_{ik} = a_{ij} +_f a_{jk} = a_{ij} + a_{jk} - 0.5 \text{ for all } i, j, k \in \{1, \cdots, n\} \ .$$

The following theorem by [10] is useful, as it shows that if G is a proper open interval of **R** and $\leq$ is the total order on G inherited from the usual order on **R**, then a continuous real alo-group can be built by the real alo-group $\mathscr{R}$, or the real alo-group $\mathscr{R}_+$. The proof of the following theorem can be found in [10].

Theorem 3.1 *Let G be a proper open interval of **R** and $\leq$ the total order on G inherited from the usual order on **R**, let $\odot$ be a binary operation on G. Then $\odot$ is a continuous, associative, and cancellative operation if and only if there exists a continuous and strictly monotonic function $\phi : J \rightarrow G$ such that or each $x, y \in J$:*

$$x \odot y = \phi(\phi^{-1}(x) + \phi^{-1}(y)) \ . \tag{3.6}$$

*and J is **R** or one of the real intervals $] - \infty; \gamma[, \ ] - \infty; \gamma], \ ]\delta; +\infty[, \ [\delta; +\infty[,$ where γ and δ are suitable constants. The function ϕ in (3.6) is unique up to a linear transformation of the variable x.*

It is clear that by (3.6) $\odot$ is commutative and strictly increasing in both variables. By Theorem 3.1 we obtain the following theorem, see [9].

Theorem 3.2 *Let G be a proper open interval of **R** and $\leq$ the total order on G inherited from the usual order on **R**. The following assertions are equivalent:*

(i) $\mathscr{G} = (G, \odot, \leq)$ is a continuous alo-group;
(ii) there exists a continuous and strictly increasing function $\phi : \mathbf{R} \rightarrow G$ verifying the equality in (3.6);

(iii) *there exists a continuous and strictly increasing function* $\psi :]0; +\infty[\to G$ *verifying the equality*

$$x \odot y = \psi(\psi^{-1}(x) \cdot \psi^{-1}(y)) \text{ for each } x, y \in]0; +\infty[. \tag{3.7}$$

Proof First, we prove the equivalence of *(i)* and *(ii)*. By Theorem 3.1, $\mathscr{G}$ is a continuous alo-group if and only if there exists a continuous and strictly monotonic function $\phi : J \to G$ defined on a proper interval J of $\mathbf{R}$, verifying (3.6). This function can be chosen as strictly increasing as it is unique up to the linear transformation of the variable x. In order to prove the required equivalence, it is enough to prove that J, the domain of ϕ, coincides with $\mathbf{R}$. By (3.6) observe that for $x \in J$:

$$x = x \odot e \Leftrightarrow \phi^{-1}(x) + \phi^{-1}(e) = 0 .$$

Hence $0 \in J$ and

$$x \odot x^{(-1)} = e \Leftrightarrow \phi^{-1}(x) + \phi^{-1}(x^{(-1)}) = \phi^{-1}(e) = 0 \Leftrightarrow \phi^{-1}(x^{(-1)}) = -\phi^{-1}(x) .$$

Therefore, if $a = \phi^{-1}(x) \in J$ then $-a = \phi^{-1}(x) \in J$. The rest follows by Theorem 3.1.

Second, we prove the equivalence of *(ii)* and *(iii)*. Let *(ii)* be true. Setting ψ as a composition of ϕ and logarithmic function log, i.e., $\psi(x) = \phi(\log x)$ for $x \in]0, +\infty[$, we obtain

$$\psi^{-1}(y) = \exp(\phi^{-1}(y)) \text{ and } \psi(\psi^{-1}(x)\psi^{-1}(y)) =$$

$$= \phi(\log(\exp(\phi^{-1}(x)) \exp(\phi^{-1}(y)))) = x \odot y .$$

Therefore, *(iii)* is true. The reverse implication can be proven in an analogous way.

$\square$

Applying Theorem 3.2, we provide, in the following propositions, two examples of continuous real alo-groups over a limited interval of $\mathbf{R}$.

Proposition 3.1 *Let* $+_f$ *be the binary operation defined for all* $x, y \in \mathbf{R}$ *by*

$$x +_f y = x + y - 0.5 \text{ for all } x, y \in \mathbf{R} . \tag{3.8}$$

and $\le$ *be the order inherited by the usual order in* $\mathbf{R}$*. Then* $\mathscr{R}_a = (\mathbf{R}, +_f, \le)$ *is a continuous alo-group with* $e = 0.5$, $x^{(-1)} = 1 - x$ *for each* $x \in \mathbf{R}$.

Proof The function $g :]0; +\infty[\to \mathbf{R}$ defined by

$$g(t) = 0.5(1 + \log t) \text{ for each } t \in]0; +\infty[, \tag{3.9}$$

is a bijection between $]0; +\infty[$ and $\mathbf{R}$ being continuous and strictly increasing. For $a, b \in]0; +\infty[$ and $x = g(a), y = g(b)$, applying (3.8) we get

$$g(a) +_f g(b) = 0.5(1 + \log a) + 0.5(1 + \log b) - 0.5 = 0.5(1 + \log a.b) = g(a \cdot b) . \quad (3.10)$$

Hence, $x +_f y = g(g^{-1}(x) \cdot g^{-1}(y))$, and (3.7) is verified with $\psi = g$. Finally, it is easy to verify that $x +_f 0.5 = x$ and $x +_f (1 - x) = 0.5$. $\qquad\square$

Proposition 3.2 *Let $\cdot_f$:]0; 1[$^2\to$]0; 1[be the operation defined for all $x, y \in$]0; 1[by*

$$x \cdot_f y = \frac{xy}{xy + (1 - x)(1 - y)} \qquad (3.11)$$

and $\leq$ be the order inherited by the usual order in **R**. *Then*]0; 1[$_m$= (]0; 1[, $\cdot_f$, $\leq$) *is a continuous alo-group with $e = 0, 5$ and $x^{(-1)} = 1 - x$ for each $x \in$]0; 1[.*

Proof The function v :]0; $+\infty$[$\to$]0; 1[defined for all $t \in$]0; $+\infty$[by

$$v(t) = \frac{t}{t + 1}$$

is a bijection between]0; $+\infty$[and]0; 1[being continuous and strictly increasing.
For $a, b \in$]0; $+\infty$[and $x = v(a), y = v(b)$, we get

$$v(a) \cdot_f v(b) = \frac{ab}{ab + 1} = v(a \cdot b) . \qquad (3.12)$$

Hence, $x \cdot_f y = v(v^{-1}(x) \cdot v^{-1}(y))$, and (3.6) is verified with $\psi = v$. Finally, it is easy to verify that $x \cdot_f 0, 5 = x$ and $x \cdot_f (1 - x) = 0, 5$. $\qquad\square$

By Examples 3.1 and 3.2 and Proposition 3.2, we shall call

- $\mathscr{R} = (\mathbf{R}, +, \leq)$ the *additive alo-group*, or *a-alo-group*;
- $\mathscr{R}_+ = (]0; +\infty[, \cdot, \leq)$ the *multiplicative alo-group*, or *m-alo-group*;
- $\mathscr{R}_a = (\mathbf{R}, +_f, \leq)$ the *fuzzy-additive alo-group*, or *fa-alo-group*;
-]0; 1[$_m$= (]0; 1[, $\cdot_f$, $\leq$) the *fuzzy-multiplicative alo-group*, or *fm-alo-group*.

Evidently, the isomorphism between $\mathscr{R}_+$ and $\mathscr{R}$ is

$$h : x \in]0; +\infty[\to \log x \in \mathbf{R}, h^{-1} : y \in \mathbf{R} \to \exp(y) \in]0; +\infty[. \qquad (3.13)$$

The isomorphism between $\mathscr{R}_+$ and $\mathbf{R}_a$ is the function g in (3.9) and its inverse, i.e.,

$$g : x \in]0; +\infty[\to 0.5(1 + \log x) \in \mathbf{R}, \ g^{-1} : y \in \mathbf{R} \to \exp(2y - 1) \in]0; +\infty[. \qquad (3.14)$$

The isomorphism between $\mathscr{R}_+$ and]0; 1[$_m$ is the function v in (3.6) and its inverse, i.e.,

$$v : x \in]0; +\infty[\to \frac{1 - x}{1 + x} \in]0; 1[, \ v^{-1} : y \in]0; 1[\to \frac{y}{1 - y} \in]0; +\infty[. \qquad (3.15)$$

3.1.3 Pairwise Comparison Matrices over a Divisible Alo-Group

Let $\mathscr{G} = (G, \odot, \leq)$ be a divisible alo-group. A pairwise comparison system over G is a pair $(\mathscr{C}, R)$ constituted by a set of alternatives $\mathscr{C} = \{c_1, c_2, \cdots, c_n\}$ and a relation $R : \mathscr{C}^2 \to G$, where $R(c_i, c_j) = a_{ij} \in G$, and $A = \{a_{ij}\}$ is a pairwise comparison matrix (PC matrix). In the context of an evaluation problem, the element a_{ij} can be interpreted as a measure on G of the preference of c_i over c_j. Here, $a_{ij} > e$ implies that c_i is strictly preferred to c_j, whereas $a_{ij} < e$ expresses the opposite and $a_{ij} = e$ means that c_i and c_j are indifferent. Moreover, $A = \{a_{ij}\}$ is assumed to be reciprocal with respect to the operation $\odot$, that is,

$$a_{ji} = a_{ij}^{(-1)} \text{ for all } i, j \in \{1, \cdots, n\}. \tag{3.16}$$

Hence, by definition $a_{ii} = e$ for all $i \in \{1, \cdots, n\}$, and $a_{ij} \odot a_{ji} = e$ for all $i, j \in \{1, \cdots, n\}$.

In the sequel, by $PC_n(\mathscr{G})$, the set of all $\odot$-reciprocal PC matrices of order $n \geq 3$ over $\mathscr{G}$ will be denoted. In particular, a matrix of $PC_n(\mathscr{R})$ is an *additive* PC matrix, a matrix of $PC_n(\mathscr{R}_+)$ is a *multiplicative* PC matrix, while a matrix of $PC_n(\mathscr{R}_a)$, or $PC_n(]0; 1[_m)$, is a *fuzzy* PC matrix.

Let $A = \{a_{ij}\} \in PC_n(\mathscr{G})$. The following notation will be used.

- A_i is the ith row of A, i.e., $A_i = (a_{i1}, a_{i2}, \cdots, a_{in})$;
- A^j is the jth column of A, i.e., $A^j = (a_{1j}, a_{2j}, \cdots, a_{nj})$;
- $m_\odot(A_i)$ is the $\odot$-mean of row $A_i = (a_{i1}, a_{i2}, \cdots, a_{in})$;
- $w_{m_\odot}(A) = (m_\odot(A_1), m_\odot(A_2), ..., m_\odot(A_n))$ is the vector of $\odot$-means of rows of A;
- $\rho_{ijk} = a_{ik} \div (a_{ij} \odot a_{jk})$ is the element of G.

By the definition of $\mathscr{G}$-distance, (see Chap. 1), we obtain

$$d_{\mathscr{G}}(a_{ik}, (a_{ij} \odot a_{jk})) = \|\rho_{ijk}\|. \tag{3.17}$$

Because of the assumption of $\odot$-reciprocity, the equality $a_{ik} = a_{ij} \odot a_{jk}$ does not depend on the considered order of the indexes i, j, k, that is,

$$a_{ik} = a_{ij} \odot a_{jk} \Leftrightarrow a_{ij} = a_{ik} \odot a_{kj} \Leftrightarrow a_{jk} = a_{ji} \odot a_{ik} \Leftrightarrow \tag{3.18}$$

$$a_{jk} = a_{ji} \odot a_{ik} \Leftrightarrow a_{ji} = a_{jk} \odot a_{ki} \Leftrightarrow \cdots. \tag{3.19}$$

Definition 3.1 Let $A = \{a_{ij}\} \in PC_n(\mathscr{G})$. A PC matrix $A = \{a_{ij}\}$ is $\odot$-*consistent* if

$$a_{ik} = a_{ij} \odot a_{jk} \text{ for all } i, j, k \in \{1, \cdots, n\}. \tag{3.20}$$

By equivalences in (3.19), we may restrict ourselves only to the cases $i < j < k$, $i, j, k \in \{1, \cdots, n\}$.

Definition 3.2 A vector $w = (w_1, w_2, \cdots, w_n)$ with $w_i \in G$ for all $i \in \{1, \cdots, n\}$ is an $\odot$-*consistent vector with respect to* $A = \{a_{ij}\} \in PC_n(\mathscr{G})$ if

$$w_i \div w_j = a_{ij} \text{ for all } i, j \in \{1, \cdots, n\} . \tag{3.21}$$

By (3.21) and the equivalences in changing the indexes (3.19), we obtain for $i, j \in \{1, \cdots, n\}$:

$$w_i > w_j \Longleftrightarrow a_{ij} > e \text{ and } w_i = w_j \Longleftrightarrow a_{ij} = e . \tag{3.22}$$

Thus, the elements of the $\odot$-consistent vector of $A = \{a_{ij}\}$ correspond with the preferences expressed by the entries a_{ij} of the PC matrix.

Proposition 3.3 $A = \{a_{ij}\} \in PC_n(\mathscr{G})$ *is* $\odot$-*consistent if and only if there exists a* $\odot$-*consistent vector* $w = (w_1, w_2, \cdots, w_n)$, $w_i \in G$.

Proof Let $A = \{a_{ij}\}$ be $\odot$-consistent. Then by (3.20), $a_{ij} = a_{ik} \odot a_{kj} = a_{ik} \div a_{jk}$ for all $i, j, k \in \{1, \cdots, n\}$. Hence, the equalities (3.21) are verified for each column vector $w = A^k$, $k \in \{1, \cdots, n\}$. On the other hand, if w is an $\odot$-consistent vector, then

$$a_{ij} \odot a_{jk} = (w_i \div w_j) \odot (w_j \div w_k) = w_i \odot w_j^{(-1)} \odot w_j \odot w_k^{(-1)}$$

$$= w_i \odot w_k^{(-1)} = a_{ik} .$$

$\square$

Proposition 3.4 *Let* $A = \{a_{ij}\} \in PC_n(\mathscr{G})$. *The following assertions are equivalent:*

(i) $A = \{a_{ij}\}$ *is* $\odot$-*consistent;*
(ii) *Each column* A^k *is an* $\odot$-*consistent vector;*
(iii) $w_{m_\odot}(A) = (m_\odot(A_1), m_\odot(A_2), \cdots, m_\odot(A_n))$ *—the vector of* $\odot$-*means of rows of* A *is an* $\odot$-*consistent vector.*

Proof The equivalence of (i) and (ii) follows directly from Proposition 3.2.
Let (ii) be true, then we get

$$(m_\odot(A_i) \div m_\odot(A_j))^{(n)} = (m_\odot(A_i))^{(n)} \div (m_\odot(A_j))^{(n)} =$$

$$= (a_{i1} \odot a_{i2} \odot \ldots \odot a_{in}) \odot (a_{j1} \odot a_{j2} \odot \cdots \odot a_{jn})^{(-1)} =$$

$$= (a_{i1} \odot a_{j1}) \odot \cdots \odot (a_{in} \odot a_{jn}) = a_{ij}^{(n)} .$$

Therefore, (3.21) is verified and (iii) is true. The opposite implication is similar. $\square$

Proposition 3.5 *Let* $\mathscr{G} = (G, \odot, \leq)$ *and* $\mathscr{G}' = (G', \circ, \preceq)$ *be divisible alo-groups and* $h : G \to G'$ *be an isomorphism between* $\mathscr{G}$ *and* $\mathscr{G}'$, $a, b \in G$, $a', b' \in G'$. *Then*

$$H : A = \{a_{ij}\} \in PC_n(\mathscr{G}) \to H(A) = \{h(a_{ij})\} \in PC_n(\mathscr{G}')$$

is a bijection between $PC_n(\mathscr{G})$ *and* $PC_n(\mathscr{G}')$ *that preserves the* $\odot$-*consistency, that is* A *is* $\odot$-*consistent if and only if* A' *is* $\circ$-*consistent.*

Proof The mapping H is an injection because h is an injective function. By applying h to the entries of the matrix $A = \{a_{ij}\}$, we get the matrix $A = \{h(a_{ij})\}$, which is reciprocal too. Therefore, $H(A) = \{h(a_{ij})\} \in PC_n(\mathscr{G}')$. The rest of the proof is straightforward. $\square$

3.2 Desirable Properties of the Priority Vector

This section generalizes the results concerning the priority vectors presented in Chap. 2. In Chap. 2 we have investigated the priority vectors associated with PCMs on the multiplicative alo-group. Here, we extend our interest to a general alo-group. Some inconsistent pairwise comparisons matrices violate the so-called fundamental selection (FS) condition. Inconsistent PCMs that violate the FS condition should not be used for the derivation of weights of alternatives and, consequently, the best alternative. The other PCMs may violate the preservation of order of preferences (the so-called POP condition), preservation of reliable preference condition (RP condition), or preservation of the intensity of preference (the so-called POIP condition), see Bana e Costa and Vansnick [2], and Mazurek and Ramik [28].

We formulate a new optimization problem for generating a priority vector (weights of alternatives, criteria, or other objects). The method is designed to find a priority vector so that all the aforementioned properties are satisfied, hence providing a more logical solution than the classical eigenvalue (EV), or the geometric mean (GM) methods.

Definition 3.3 Let $A = \{a_{ij}\} \in PC_n(\mathscr{G})$ be the PC matrix associated with the set of alternatives $\mathscr{C} = \{c_1, c_2, \cdots, c_n\}$ with elements a_{ij} from alo-group $\mathscr{G}$. We say that an alternative c_i *dominates alternative* c_j or, equivalently, that an alternative c_j *is dominated by alternative* c_i, if

$$a_{ij} > e, \tag{3.23}$$

where e is the identity element of $\mathscr{G}$. If a given alternative is not dominated by any other alternative, then such alternative is called the *non-dominated alternative*. The set of all non-dominated alternatives in $\mathscr{C}$ with respect to matrix A is denoted by $ND(A)$. By (3.23) we obtain

$$ND(A) = \{c_j \in \mathscr{C} |\ \text{there is no } i \in \{1, \cdots, n\} : a_{ij} > e\}. \tag{3.24}$$

We say that the *fundamental selection (FS) condition* is satisfied with respect to $A \in PC_n(\mathcal{G})$ and $w \in G^n$, if the maximal weight of the priority vector is associated with a non-dominated alternative.

Equivalently, we say that w satisfies the FS condition with respect to A, if for some $i^* \in \{1, \cdots, n\}$

$$c_{i^*} \in ND(A) \text{ and } w(c_{i^*}) = \max\{w(c_j)| j \in \{1, \cdots, n\}\}. \tag{3.25}$$

We also say that A satisfies the FS condition with respect to w.

In [2, 28], the conditions of order preservations are formulated in multiplicative system. Here, we formulate these conditions in a more general setting, i.e., in alo-groups. The first, *the preservation of order preference condition (POP)*, claims that the ranking result in relation to the given pair of alternatives (a_i, a_j) should not break with the expert judgment.

A PC matrix $A \in PC_n(\mathcal{G})$ is said to satisfy the *preservation of order preference condition (POP condition) with respect to priority vector w* if

$$a_{ij} > e \Rightarrow w_i > w_j. \tag{3.26}$$

A PC matrix A is said to satisfy the *preservation of order intensity preference condition (POIP condition) with respect to vector w* if

$$a_{ij} > e, a_{kl} > e, \text{ and } a_{ij} > a_{kl} \Rightarrow w_i \div w_j > w_k \div w_l. \tag{3.27}$$

A PC matrix A is said to satisfy the *reliable preference (RP) condition with respect to priority vector w* if

$$a_{ij} > e \Rightarrow w_i > w_j, \tag{3.28}$$

$$a_{ij} = e \Rightarrow w_i = w_j. \tag{3.29}$$

Remark 3.1 From (3.28) in the above definition it will be shown that if a PC matrix A satisfies the RP condition with respect to priority vector w, then A satisfies the POP condition with respect to priority vector w. The opposite is not true, [26].

Proposition 3.6 *Let $A = \{a_{ij}\} \in PC_n(\mathcal{G})$, and let $w = (w_1, \cdots, w_n)$ be a priority vector associated with A. If A satisfies the POP condition with respect to w, then A satisfies the FS condition with respect to w.*

Proof Assume that c_{i_0} is non-dominated, and w_{j_0} is the maximal weight, $i_0 \neq j_0$. If c_{j_0} is non-dominated, then the proposition holds. On the other hand, if c_{j_0} is dominated, then there is a c_{k_0} which dominates c_{j_0}, hence $a_{k_0 j_0} > e$. By the POP condition we have $w_{k_0} > w_{j_0}$, a contradiction with the assumption that w_{j_0} is a maximal weight.

Hence, the FS condition is satisfied. $\square$

Definition 3.4 Let $A = \{a_{ij}\} \in PC_n(\mathcal{G})$. For each pair $i, j \in \{1, \cdots, n\}$, and a priority vector $w = (w(c_1), \cdots, w(c_n))$ the *error element*, $\varepsilon(i, j, w)$, is given as

$$\varepsilon(i, j, w) = a_{ij} \odot w(c_j) \div w(c_i), \tag{3.30}$$

and similarly (as in [25]) let us define

$$\mathcal{E}(i, j, w) = \max\{\varepsilon(i, j, w), (\varepsilon(i, j, w))^{(-1)}\}. \tag{3.31}$$

The *error index*, $\mathcal{E}(A, w)$, for the PC matrix A and a priority vector $w = (w(c_1), \cdots , w(c_n))$, is the maximal value of $\mathcal{E}(i, j, w)$, i.e.,

$$\mathcal{E}(A, w) = \max_{i,j \in \{1, \cdots , n\}} \mathcal{E}(i, j, w). \tag{3.32}$$

The *minimal error index*, $\mathcal{E}(A)$, for the PC matrix A is the infimum of $\mathcal{E}(A, w)$ for all (normalized) priority vectors $w = (w_1, \cdots , w_n)$.

Remark 3.2 In our effort to set up the "best" priority vector corresponding to the PC matrix A, we solve the following optimization problem (P):

(P)

$$\mathcal{E}(A, w) \longrightarrow \min; \tag{3.33}$$

subject to

$$\bigodot_{k=1}^{n} w(c_k) = e, \tag{3.34}$$

$$w = (w(c_1), \cdots , w(c_n)),$$

$$w(c_k) \in G \text{ for all } k \in \{1, \cdots , n\}. \tag{3.35}$$

If there exists an optimal solution of (P), i.e., a priority vector $w^* = (w_1^*, \cdots , w_n^*)$, then we denote

$$\mathcal{E}(A) = \mathcal{E}(A, w^*). \tag{3.36}$$

Here, priority vector w^* corresponds to the minimal error index $\mathcal{E}(A)$. This situation may happen if the set of feasible solutions of (P) is bounded, then by continuity of operation $\odot$ there always exists an optimal solution of (P).

Unfortunately, the optimal solution of (P) need not exist at all, e.g., by the unboundedness of the set of feasible solutions of (P). Then an approximate solution of (P) with the value of the objective function as close as possible to $\mathcal{E}(A)$ should be found.

Now, we derive some properties of $\mathcal{E}(A, w)$.

Lemma 3.1 *Let $A = \{a_{ij}\} \in PC_n(\mathcal{G})$ and $w = (w(c_1), \cdots , w(c_n))$ be a priority vector. Then*

$$\mathcal{E}(A, w) \geq e. \qquad\qquad (3.37)$$

Moreover, if

$$\mathcal{E}(A, w) = e,$$

then A is $\odot$-consistent and

$$\mathcal{E}(A, w) = \mathcal{E}(A) = e.$$

Proof Either $\varepsilon(i, j, w) \geq e$, or $\varepsilon(i, j, w) \leq e$, then $\varepsilon(i, j, w))^{(-1)} \geq e$. Hence,

$$\mathcal{E}(i, j, w) = \max\{\varepsilon(i, j, w), (\varepsilon(i, j, w))^{(-1)}\} \geq e.$$

By (3.32) we obtain

$$\mathcal{E}(A, w) \geq e.$$

Moreover, let $\mathcal{E}(A, w) = e$. Then by (3.31), (3.32) for all $i, j \in \{1, \cdots , n\}$, it holds that

$$\varepsilon(i, j, w) = a_{ij} \odot w(c_j) \div w(c_i) = e.$$

Hence, equivalently,

$$a_{ij} = w(c_i) \div w(c_j).$$

Then we obtain

$$a_{ij} \odot a_{jk} \odot a_{ki} = w(c_i) \div w(c_j) \odot w(c_j) \div w(c_k) \odot w(c_k) \div w(c_i) = e.$$

Therefore, by (3.20), A is $\odot$-consistent. The rest of the lemma follows from (3.36).
□

Remark 3.3 The error index $\mathcal{E}(A, w)$ depends not only on the elements a_{ij} of PC matrix A, but also on the priority vector w. It is, however, always greater than or equal to the identity element $e \in G$. If the error index of A is equal to e then PC matrix A is $\odot$-consistent. Later on, in Lemma 3.2, we will show that for an $\odot$-consistent PC matrix A there is always a priority vector w such that $\mathcal{E}(A, w) = e$ holds.

Definition 3.5 Let $A = \{a_{ij}\} \in PC_n(\mathcal{G})$. For the pair of alternatives $c_i, c_j \in \mathcal{C}$ such that c_i dominates c_j (i.e., $a_{ij} > e$) and a priority vector $w = (w(c_1), \cdots , w(c_n))$, it holds that

$$w(c_i) > w(c_j), \text{ or equivalently } w(c_i) \div w(c_j) > e. \qquad (3.38)$$

Then we say that *the preservation of order preference condition (POP)* is satisfied for w in relation to the given PC matrix A.

Definition 3.6 Let $A = \{a_{ij}\} \in PC_n(\mathcal{G})$. For the pairs of alternatives $c_i, c_j \in \mathcal{C}$ and $c_k, c_l \in \mathcal{C}$ such that $a_{ij} > e$, $a_{kl} > e$ and $a_{ij} > a_{kl}$, and a priority vector $w = (w(c_1), ..., w(c_n))$, it holds

$$w(c_i) \div w(c_j) > w(c_k) \div w(c_l). \tag{3.39}$$

Then *the preservation of order of intensity preference condition (POIP)* is satisfied for w in relation to the given PC matrix A.

We start by showing that the POP and POIP conditions are satisfied if the PC matrix is $\odot$-consistent. This is the well-known necessary and sufficient condition for a PC matrix to be $\odot$-consistent (see also [26, 32]).

Lemma 3.2 *Let $A = \{a_{ij}\} \in PC_n(\mathcal{G})$. Then A is $\odot$-consistent if and only if there exists a priority vector $w = (w(c_1), \cdots, w(c_n))$ such that for all $i, j \in \{1, \cdots, n\}$*

$$w(c_i) \div w(c_j) = a_{ij}. \tag{3.40}$$

Proof Suppose that $A = \{a_{ij}\}$ is $\odot$-consistent, then by (3.20)

$$a_{ij} \odot a_{jk} \odot a_{ki} = e \text{ for all } i, j, k \in \{1, \cdots, n\},$$

or, equivalently,

$$a_{ij} \odot a_{jk} = a_{ik}.$$

Let $w = (w(c_1), \cdots, w(c_n))$ be given by "generalized means" as

$$w(c_i) = \left(\bigodot_{r=1}^{n} a_{ir}\right)^{\left(\frac{1}{n}\right)}, i \in \{1, \cdots, n\}, \tag{3.41}$$

Then we obtain the consistency condition $a_{ir} \odot a_{rj} = a_{ij}$

$$w(c_i) \div w(c_j) = \left(\bigodot_{r=1}^{n} a_{ir}\right)^{\left(\frac{1}{n}\right)} \odot \left(\left(\bigodot_{r=1}^{n} a_{jr}\right)^{\left(\frac{1}{n}\right)}\right)^{(-1)} =$$

$$= \left(\bigodot_{r=1}^{n}(a_{ir} \odot a_{rj})\right)^{\left(\frac{1}{n}\right)} = \left(a_{ij}^{(n)}\right)^{\left(\frac{1}{n}\right)} = a_{ij}.$$

Hence, (3.40) is satisfied. On the other hand, let condition (3.40) be true. Then for each $i, j, k \in \{1, \cdots, n\}$ we obtain

$$a_{ij} \odot a_{jk} \odot a_{ki} = (w(c_i) \div w(c_j)) \odot (w(c_j) \div w(c_k)) \odot (w(c_k) \div w(c_i)) =$$

$$= w(c_i) \odot w(c_j)^{(-1)} \odot w(c_j) \odot w(c_k)^{(-1)} \odot w(c_k) \odot w(c_i)^{(-1)} = e.$$

Hence, A is $\odot$-consistent. $\qquad\square$

Theorem 3.3 *Let $A = \{a_{ij}\} \in PC_n(\mathscr{G})$. If A is $\odot$-consistent then there exists a priority vector $w = (w(c_1), \cdots , w(c_n))$ such that for all $i, j \in \{1, \cdots , n\}$, the POP condition is satisfied, i.e., $a_{ij} > e$ implies $w_i > w_j$.*

The RP condition is also satisfied.

Moreover, if $a_{kl} > e$, $k, l \in \{1, \cdots , n\}$, then the POIP condition is also satisfied, i.e., $a_{ij} > a_{kl}$ implies that $w_i \div w_j > w_k \div w_l$.

Proof Suppose that $A = \{a_{ij}\}$ is $\odot$-consistent. If for some $i, j \in \{1, \cdots , n\}$ we have $a_{ij} > e$, then by Lemma 3.2 there exists $w = (w(c_1), \cdots , w(c_n))$ such that

$$a_{ij} = w(c_i) \div w(c_j) > e,$$

which is equivalent to $w(c_i) > w(c_j)$ and the POP condition is satisfied.

If $a_{ij} = w(c_i) \div w(c_j) = e$, then $w(c_i) = w(c_j)$.

Moreover, by Lemma 3.2, it holds that $a_{ij} > a_{kl}$ if and only if

$$w(c_i) \div w(c_j) > w(c_k) \div w(c_l),$$

hence, (3.39) is satisfied. $\qquad\square$

Lemma 3.3 *Let $A = \{a_{ij}\} \in PC_n(\mathscr{G})$ and $w = (w(c_1), \cdots , w(c_n))$ be a priority vector. Then for all $i, j \in \{1, \cdots , n\}$*

$$\mathscr{E}(A, w)^{(-1)} \odot w(c_i) \div w(c_j) \le a_{ij} \le \mathscr{E}(A, w) \odot w(c_i) \div w(c_j). \tag{3.42}$$

Proof By (2.42), (3.32) we obtain

$$\mathscr{E}(A, w) \ge \max\{\varepsilon(i, j, w), (\varepsilon(i, j, w))^{(-1)}\} \ge \varepsilon(i, j, w) = a_{ij} \odot w(c_j) \div w(c_i), \tag{3.43}$$

$$\mathscr{E}(A, w) \ge \max\{\varepsilon(i, j, w), (\varepsilon(i, j, w)^{(-1)}\} \ge \varepsilon(i, j, w)^{(-1)} = a_{ji} \odot w(c_i) \div w(c_j). \tag{3.44}$$

Hence, when multiplying both sides of (3.43) by $w(c_i) \div w(c_j)$, and both sides of (3.44) by $w(c_j) \div w(c_i)$, we get

$$\mathscr{E}(A, w) \odot w(c_i) \div w(c_j) \ge a_{ji} \odot w(c_i) \div w(c_j) \odot w(c_j) \div w(c_i) = a_{ij} \odot e = a_{ij}. \tag{3.45}$$

$$\mathscr{E}(A, w)^{(-1)} \odot w(c_i) \div w(c_j) \le a_{ij} \odot w(c_j) \div w(c_i) \odot w(c_i) \div w(c_j) = a_{ij} \odot e = a_{ij}. \tag{3.46}$$

Combining (3.45) and (3.46) we obtain (3.42). $\qquad\square$

Now, we turn our attention to $\odot$-inconsistent $\odot$-reciprocal PC matrices. The following theorem gives sufficient conditions for validity of the POP and POIP.

Theorem 3.4 *Let $A = \{a_{ij}\} \in PC_n(\mathcal{G})$, and let $w = (w(c_1), \cdots, w(c_n))$ be a priority vector, let $i, j, k, l \in \{1, \cdots, n\}$.*
  If

$$a_{ij} > \mathcal{E}(A, w),$$

then

$$w(c_i) > w(c_j).$$

Moreover, if

$$a_{kl} > \mathcal{E}(A, w)$$

and

$$a_{ij} \div a_{kl} > (\mathcal{E}(A, w))^2, \tag{3.47}$$

then

$$w(c_k) > w(c_l)$$

and

$$w(c_i) \div w(c_j) > w(c_k) \div w(c_l).$$

Hence, the POP and also POIP condition is satisfied for w in relation to the corresponding pairs of alternatives.

Proof If for some $i, j \in \{1, \cdots, n\}$ we have $a_{ij} > \mathcal{E}(A, w)$, then by (3.42) in Lemma 3.3 and $a_{ij} > \mathcal{E}(A, w)$ we obtain

$$\mathcal{E}(A, w) \odot w(c_i) \div w(c_j) \geq a_{ij} > \mathcal{E}(A, w), \tag{3.48}$$

which implies, when both sides of (3.48) are "multiplied" by $\mathcal{E}(A, w)^{(-1)}$,

$$w(c_i) \div w(c_j) > e,$$

i.e., $w_i > w_j$, the POP condition is satisfied. Similarly, if for some $k, l \in \{1, \cdots, n\}$ we have $a_{ij} > \mathcal{E}(A, w), a_{kl} > \mathcal{E}(w)$, then we obtain $w(c_i) > w(c_j)$ and $w(c_k) > w(c_l)$, i.e., POP condition. Moreover, by (3.42) in Lemma 3.3 we obtain

$$\mathcal{E}(A, w) \odot w(c_i) \div w(c_j) \geq a_{ij}, \tag{3.49}$$

$$\mathcal{E}(A, w) \odot w(c_l) \div w(c_k) \geq a_{lk} = a_{kl}^{(-1)}. \tag{3.50}$$

By "$\odot$-multiplying" the left side and right side of (3.48) and (3.50), we obtain

$$(\mathcal{E}(A, w))^2 \odot (w(c_i) \div w(c_j)) \odot (w(c_k) \div w(c_l))^{(-1)} \geq a_{ij} \div a_{kl}. \tag{3.51}$$

If we assume that $(w(c_i) \div w(c_j)) \odot (w(c_k) \div w(c_l))^{(-1)} \le e$, which is equivalent to $w(c_i) \div w(c_j) \le w(c_k) \div w(c_l)$, then by (3.51) we obtain

$$(\mathscr{E}(A, w))^2 \ge a_{ij} \div a_{kl}.$$

This result, however, is in contradiction with (3.47). Hence, it must be $w(c_i) \div w(c_j) > w(c_k) \div w(c_l)$. Therefore, the POIP condition is satisfied. $\square$

Remark 3.4 Notice that in the previous lemmas and theorems, there is no special assumption concerning the method for generating the priority vector $w = (w(c_1), \cdots , w(c_n))$. The priority vector or vector of weights w may be an arbitrary vector with normalized elements from alo-group G. Specifically, in the case of a multiplicative alo-group of positive real numbers $\mathscr{R}_+ = (\mathbf{R}_+, \cdot, \le)$ with some field operation +, we may use *EVM*, *GMM* or any other priority vector generating method.

In the following section, we shall deal with an optimization method satisfying our POP, RP, and POIP conditions and minimizing at the same time the error index.

3.3   Deriving Priority Vector by Solving an Optimization Problem

Let $A = \{a_{ij}\} \in PC_n(\mathscr{G})$ be a PC matrix. Based on this PCM, we define the following three sets of indexes:

$$I^{(1)}(A) = \{(i, j) | i, j \in \{1, \cdots , n\}, a_{ij} = e\}, \tag{3.52}$$

$$I^{(2)}(A) = \{(i, j) | i, j \in \{1, \cdots , n\}, a_{ij} > e\}, \tag{3.53}$$

$$I^{(4)}(A) = \{(i, j, k, l) | i, j, k, l \in \{1, \cdots , n\}, a_{ij} > e, a_{kl} > e, a_{ij} > a_{kl}\}. \tag{3.54}$$

An *error index*, $\mathscr{E}(A, w)$, of $A = \{a_{ij}\}$ and $w = (w_1, \cdots , w_n)$ has been defined as follows:

$$\mathscr{E}(A, w) = \max_{\{i,j \in \{1,\cdots ,n\}\}} \{\|a_{ij} \odot w_j \div w_i\|\}. \tag{3.55}$$

The problem of finding a priority vector satisfying the FS, POP, RP, and POIP conditions can be formulated in terms of the following optimization problem, where $A = \{a_{ij}\}$ is a given PC matrix and $w = (w_1, \cdots , w_n)$ is an unknown priority vector with variables $w_1, \cdots , w_n$:

(⊙-**Problem 0**)

$$\mathscr{E}(A, w) \longrightarrow \min; \tag{3.56}$$

subject to

$$\bigodot_{r=1}^{n} w_r = e, w_r > e \ \forall r, \tag{3.57}$$

$$w_r = w_s \ \forall (r, s) \in I^{(1)}(A), \tag{3.58}$$

$$w_r > w_s \ \forall (r, s) \in I^{(2)}(A), \tag{3.59}$$

$$w_r \div w_s > w_t \div w_u \ \forall (r, s, t, u) \in I^{(4)}(A). \tag{3.60}$$

If $w^* = (w_1^*, \cdots, w_n^*)$ is an optimal solution of ($\odot$- Problem 0), then the objective function in (3.56) minimizes the error index of PC matrix A. By constraint (3.57), the weights are positive and ($\odot$)-normalized. By (3.58), resp. (3.59), the RP, resp. POP condition is secured and by (3.60) the POIP condition is met.

3.3.1 Transformation to ($\odot$-Problem ε)

Unfortunately, ($\odot$- Problem 0) is not in a standard form of optimization problem that is appropriate for solving by existing numerical methods. Here, variables w_i are required to be strictly positive and some inequality constraints, (3.59), (3.60), are strict, so the set of feasible solution is not closed. That is why we transform the problem into a more convenient form. Given $\varepsilon > e$.

($\odot$- **Problem** ε)

$$\mathscr{E}(A, w) \longrightarrow \min; \tag{3.61}$$

subject to

$$\bigodot_{r=1}^{n} w_r = e, w_r \geq \varepsilon \ \forall r, \tag{3.62}$$

$$w_r = w_s \ \forall (r, s) \in I^{(1)}(A), \tag{3.63}$$

$$w_r \geq w_s \odot \varepsilon \ \forall (r, s) \in I^{(2)}(A), \tag{3.64}$$

$$w_r \div w_s \geq w_t \div w_u \odot \varepsilon \ \forall (r, s, t, u) \in I^{(4)}(A). \tag{3.65}$$

Notice that here, strict inequalities have been changed to non-strict ones here by including a sufficient constant $\varepsilon > e$.

The following proposition says that both problems, i.e., ($\odot$-Problem 0) and ($\odot$-Problem ε), are in some sense equivalent. The proof of the proposition is evident.

Proposition 3.7 ($\odot$-Problem 0) *has a feasible solution w if and only if there exists* $\varepsilon > e$ *such that w is a feasible solution of* ($\odot$-Problem ε).

Moreover, if ($\odot$-Problem 0) has an optimal solution w^* then there exists $\varepsilon > e$ such that w^* is an optimal solution of ($\odot$-Problem ε).

Here, the problem ($\odot$-Problem ε) will be decomposed into the following three optimization problems depending on the particular formulation of the objective function (3.61) as well as constraints (3.62)–(3.65), i.e., nested sets of feasible solutions. We shall consider the following optimization problem variants:

(I) Minimize the objective function (3.61), subject to (3.62). The optimal solution is denoted by $w^{(I)}$. The FS, POP, RP and POIP conditions are not necessarily satisfied.

(II) Minimize the objective function (3.61) subject to constraints (3.62)–(3.64). The optimal solution is denoted by $w^{(II)}$. The POP and RP conditions are satisfied; then by Lemma 3.2 the FS condition is also satisfied. The POIP condition is not necessarily satisfied.

(III) Minimize the objective function (3.61) subject to constraints (3.62)–(3.65). The optimal solution is denoted by $w^{(III)}$. Here, the FS, POP, RP, and POIP conditions are satisfied.

3.3.2 Solving ($\odot$-Problem ε)

Notice that the set of feasible solutions of ($\odot$-Problem ε), (3.61)–(3.65), could be empty, e.g., for problems (II), and/or (III). Even for a nonempty set of feasible solutions of ($\odot$-Problem 0), the optimal solution of the corresponding optimization problems (I), (II), or (III) need not exist, as the set of feasible solutions is not secured to be closed and/or bounded and the objective function need not be convex.

On the other hand, if the optimal solution $w^* = (w_1^*, \cdots, w_n^*)$ of some problems of (I)–(III) of ($\odot$-Problem ε) exists, the FS, RP, POP, and POIP conditions hold by the nested properties of the feasible solution sets. Then, $w^* = (w_1^*, \cdots, w_n^*)$ is an appropriate priority vector associated with A satisfying the required properties.

Proposition 3.8 *Let $A = \{a_{ij}\} \in PC_n(\mathcal{G})$ be an $\odot$-consistent pairwise comparison matrix. Then there is a unique optimal solution $w^* = (w_1^*, \cdots, w_n^*)$ of ($\odot$-Problem 0) satisfying*

$$a_{ij} = w_i^* \div w_j^* \ \forall i, j, \tag{3.66}$$

such that the FS, POP, RP, and POIP conditions are met.

Proof By Proposition 3.7 there exists a unique priority vector with positive components $w^* = (w_1^*, \cdots, w_n^*)$, such that

$$a_{ij} = w_i^* \div w_j^* \ \forall i, j. \tag{3.67}$$

Then $\mathscr{E}(A, w^*) = e$ and by Lemma 3.1 it is the minimal value of the objective function (3.61). Hence, $w^* = (w_1^*, \cdots, w_n^*)$ is an optimal solution of ($\odot$-Problem 0). By (3.66) we have that the FS, RP, POP, and POIP conditions are met. $\qquad\square$

When solving a particular optimization problem, ($\odot$-Problem ε), (3.61)–(3.65) with the objective function (3.61), we encounter numerical difficulties, as this optimization problem is nonlinear and also non-convex. Non-convexity is found in objective function (3.61) and also in constraints (3.65). Fortunately, these obstacles can be avoided by a proper approach—transformation of the non-convex problem to a convex one, which enables using standard numerical methods to be used for solving NLP problems. Then, for variants (I) and (II) of ($\odot$-Problem ε), we obtain an optimization problem that is solvable, for example, by efficient interior point methods (see e.g., [6]). For solving variant (III) with non-convex constraints (3.65), we can apply, e.g., an interior or exterior penalty method by penalizing this constraint and moving it into the objective function (see e.g., [6]).

3.4 Generalized Geometric Mean Method (GGMM)

Following the geometric mean method (GMM), we define the *generalized geometric mean method (GGMM)*, where the weight of the ith alternative is given by the $\odot$-mean of the ith row of $A = \{a_{ij}\}$. The question is, however, whether these weights satisfy the desired conditions (POP, RP, POIP).

Definition 3.7 Let $A = \{a_{ij}\} \in PC_n(\mathscr{G})$. A priority vector

$$w_{GGM} = (w(c_1), \cdots, w(c_n)) \tag{3.68}$$

is defined as

$$w_{GGM} = \left(\left(\bigodot_{r=1}^{n} a_{1r} \right)^{(\frac{1}{n})}, \cdots, \left(\bigodot_{r=1}^{n} a_{nr} \right)^{(\frac{1}{n})} \right). \tag{3.69}$$

The individual weights of the priority vector w_{GGM} are generated by the *generalized geometric mean method (GGMM)*. They are given as

$$w(c_i) = \left(\bigodot_{r=1}^{n} a_{ir} \right)^{(\frac{1}{n})}, i \in \{1, \cdots, n\}. \tag{3.70}$$

Remark 3.5 Notice that individual weights of the priority vector w_{GGM} are $\odot$-normalized, i.e., $\bigodot_{k=1}^{n} w(c_k) = e$, which is a consequence of reciprocity of A.

Definition 3.8 Let $A = \{a_{ij}\} \in PC_n(\mathscr{G})$. Let for $i, j, k \in \{1, \cdots, n\}$

$$e(i, j, k) = a_{ij} \odot a_{jk} \odot a_{ki}, \tag{3.71}$$

and similarly let us define

$$\eta(i, j, k) = \max\{e(i, j, k), (e(i, j, k))^{(-1)}\}. \tag{3.72}$$

The *generalized inconsistency index ($GI_\odot$) of the PC matrix A* is defined as

$$GI_\odot(A) = \max\{\eta(i, j, k) | i, j, k \in \{1, \cdots, n\}\}. \tag{3.73}$$

Lemma 3.4 *Let $A = \{a_{ij}\} \in PC_n(\mathscr{G})$, $w_{GM} = (w_{GM}(c_1), \cdots, w_{GM}(c_n))$ be a priority vector generated by GGMM (3.70).*

Then the error index $\mathscr{E}(A, w_{GM})$ is less than or equal to the generalized inconsistency index, i.e.

$$\mathscr{E}(A, w_{GM}) \leq GI(A). \tag{3.74}$$

Proof Providing the use of GGMM we have for each $i, j \in \{1, ..., n\}$

$$a_{ij} \odot w_{GM}(c_j) \div w_{GM}(c_i) = a_{ij} \odot \left(\bigodot_{k=1}^{n} a_{jk} \div \bigodot_{k=1}^{n} a_{ik} \right)^{(\frac{1}{n})}$$

thus,

$$a_{ij} \odot w_{GM}(c_j) \div w_{GM}(c_i) = \left(\bigodot_{k=1}^{n} a_{ij} \odot a_{jk} \div \bigodot_{k=1}^{n} a_{ik} \right)^{(\frac{1}{n})} = \left(\bigodot_{k=1}^{n} a_{ij} \odot a_{jk} \odot a_{ki} \right)^{(\frac{1}{n})}$$

However, it holds that

$$\left(\bigodot_{k=1}^{n} a_{ij} \odot a_{jk} \odot a_{ki} \right)^{(\frac{1}{n})} \leq \max_{k \in \{1, \cdots, n\}} \{a_{ij} \odot a_{jk} \odot a_{ki}\} =$$

$$\max_{k \in \{1, \cdots, n\}} \{\eta(i, j, k)\}, \tag{3.75}$$

hence, $\mathscr{E}(A, w_{GM}) \leq GI(A)$. $\square$

Remark 3.6 The following theorem is a reformulation of Theorem 3.4. The advantage of this reformulation is that the error index $\mathscr{E}(A, w_{GM})$ need not be calculated, instead, the generalized inconsistency index (GII), which is independent of the priority vector is necessary. Moreover, Theorem 3.5 gives a sufficient condition for the priority vector generated by the GGMM satisfying the POP and POIP conditions, see also Example 3.5.

Theorem 3.5 *Let $A = \{a_{ij}\} \in PC_n(\mathscr{G})$, let $w_{GM} = (w_{GM}(c_1), \cdots, w_{GM}(c_n))$ be a priority vector generated by GGMM (3.70). If*

$$a_{ij} > GI(A),$$

then
$$w_{GM}(c_i) > w_{GM}(c_j).$$

Moreover, if
$$a_{kl} > GI(A)$$

and
$$a_{ij} \div a_{kl} > (GI(A))^2, \tag{3.76}$$

then
$$w_{GM}(c_k) > w_{GM}(c_l)$$

and
$$w_{GM}(c_i) \div w_{GM}(c_j) > w_{GM}(c_k) \div w_{GM}(c_l).$$

Therefore, the POP and also POIP condition are satisfied for w_{GM} in relation to the PCM A.

Proof This is an immediate conclusion coming from Theorem 3.4 and Lemma 3.4.

$\square$

There exists a relationship between *Koczkodaj's* inconsistency index (2.42) and the *generalized inconsistency index* (3.73) in the multiplicative alo-group $\mathscr{R}_+ = (\mathbf{R}_+, \cdot, \leq)$ together with the additional field operation $+$, see also Chap. 2, Propositsion 2.13.

Remark 3.7 By comparing (2.38) and (3.32), (3.73) we easily derive the relation between $KI(A)$ and $GI(A)$ as follows:

$$GI(A) = \frac{1}{1 - KI(A)}. \tag{3.77}$$

Thus, the above Theorem 3.5 can also be formulated using Koczkodaj's inconsistency index (2.38).

Remark 3.8 From the point of view of practical application the following theorem seems to be important. Again, it is a reformulation of Theorem 3.4 using the optimal solution of problem (P) as the priority vector. The advantage is that then the error index $\mathscr{E}(A, w)$ attains its minimal value $\mathscr{E}(A)$ and (3.78) can be easily satisfied. We shall call this method the *minimal error index method (MEIM)*.

Theorem 3.6 *Let $A = \{a_{ij}\} \in PC_n(\mathscr{G})$, let $w^* = (w^*(c_1), \cdots, w^*(c_n))$ be a priority vector generated by the optimal solution of problem ($\odot$-Problem 0), i.e., (3.56), (3.57)–(3.60). If*
$$a_{ij} > \mathscr{E}(A), \tag{3.78}$$

then

$$w^*(c_i) > w^*(c_j).$$

Moreover, if

$$a_{kl} > \mathscr{E}(A)$$

and

$$a_{ij} \div a_{kl} > (\mathscr{E}(A))^2 , \tag{3.79}$$

then

$$w^*(c_k) > w^*(c_l)$$

and

$$w^*(c_i) \div w^*(c_j) > w^*(c_k) \div w^*(c_l).$$

Therefore, the POP, RP, and also POIP conditions are satisfied for w^* in relation to the corresponding PCM A.

Example 3.5 We consider an illustrative example of ranking $n = 4$ alternatives c_1, c_2, c_3, c_4, where information about the preferences of the DM between pairs of alternatives is given by 4×4 *PC* matrix A in the usual multiplicative alo-group of positive real numbers $\mathscr{R}_+ = (\mathbf{R}_+, \cdot, \leq)$ as follows:

$$A = \begin{pmatrix} 1 & 3 & 3 & 2 \\ \frac{1}{3} & 1 & 9 & 9 \\ \frac{1}{3} & \frac{1}{9} & 1 & 3 \\ \frac{1}{2} & \frac{1}{9} & \frac{1}{3} & 1 \end{pmatrix}.$$

Here, the PC matrix A is characterized by Koczkodaj's inconsistency index $KI(A) = 0.926$, or the Generalized inconsistency index $GI(A) = 13.500$. Notice that alternative c_1 is non-dominated.

Now, we generate the priority vector of A by three methods: GGMM, EVM (i.e., Saaty's method) and MEIM (i.e., the minimal error index method).

The priority vector w_{GGM} generated by *GGMM*, i.e., by (3.69), is:

$$w_{GGM} = (2.060, 2.280, 0.577, 0.369),$$

hence the corresponding ranking of the alternatives is $a_2 > a_1 > a_3 > a_4$. The "best" alternative is a_2. Notice that the best alternative is not non-dominated. Moreover, the error index for w_{GGM} is $\mathscr{E}(A, w_{GGM}) = 3.320$.

We conclude that if $a_{ij} > 3.320$, then $w(a_i) > w(a_j)$. By Theorem 3.2, it is clear that the POP condition holds for elements $a_{23} = a_{24} = 9$, and POIP condition is not satisfied.

The priority vector w_{EVM} generated by *EVM method*, is

$$w_{EVM} = (2.232, 2.217, 0.529, 0.382),$$

hence the corresponding ranking of the alternatives is $a_1 > a_2 > a_3 > a_4$. The "best" alternative is a_1. The error index for w_{EVM} is $\mathscr{E}(A, w_{EVM}) = 2.980$, moreover, the consistency index $CI(A) = 0.344$.

We conclude that if $a_{ij} > 2.980$, then $w(a_i) > w(a_j)$. By Theorem 2, it is clear that the POP condition holds for elements $a_{23} = a_{24} = 9$, and POIP condition is not satisfied.

The priority vector w^* generated by the optimal solution of problem (P), i.e., by MEIM, is

$$w^* = (2.246, \ 1.889, \ 0.529, \ 0.445).$$

Hence, the corresponding ranking of the alternatives is $a_1 > a_2 > a_3 > a_4$. The "best" alternative is again a_1. Moreover, the error index, i.e., minimal error index, for w^* is $\mathscr{E}(A, w^*) = \mathscr{E}(A) = 2.523$.

By Theorem 3.2, and also Theorem 3.4, it is clear, that the POP condition holds for all pairs of alternatives except (a_1, a_4) in relation to w^*. Again, the POIP condition is not satisfied.

3.5 Measuring Consistency of PCM in Alo-Groups

3.5.1 Multiplicative Alo-Group

Consider first a multiplicative system, i.e., multiplicative alo-group $\mathscr{G}$. Multiplicative consistent matrices correspond to the ideal situation in which there are the exact values $v_1, \cdots, v_n$ for the alternatives $c_1, \cdots, c_n$. Then the quotients $\{\frac{v_i}{v_j}\}$ form a consistent matrix. The vector $v = (v_1, \cdots, v_n)$ is unique up to a multiplicative constant. The challenge of the pairwise comparisons method comes from the lack of consistent PCMs which arise in practice.

Given an inconsistent $n \times n$ PC matrix $A = \{a_{ij}\}$. An approximation of a PC matrix is sufficiently good if the inconsistency is "acceptable". It can be done by localizing the inconsistency and, eventually, reducing it to some predefined limit. The inconsistency threshold is then arbitrary, usually set up by a DM. For a given inconsistent matrix A, we shall approximate it with a consistent matrix A^{con}, such that the "distance" from matrix A is "as small as possible".

The most important question: "Where does the inconsistency come from?" can be explained by the natural variability of input data. The superfluous data comes from collecting/evaluating PCM elements for all pairs combinations which is $n(n-1)/2$, while only $n-1$ comparisons is sufficient, for other details see Chap. 2, Sect. 2.5.

According to Saaty, [37], an m-inconsistency grade of a positive m-reciprocal $n \times n$ matrix $A = \{a_{ij}\}$ can be measured by the m-EV-consistency index $I_{mEV}(A)$ defined as

$$I_{mEV}(A) = \frac{\rho(A) - n}{n - 1}, \tag{3.80}$$

where $\rho(A)$ is the spectral radius of A (i.e., the principal eigenvalue of A), see also Chap. 2, Sect. 2.4. The proof of the following result can be found in [37].

Proposition 3.9 *If $A = \{a_{ij}\}$ is an $n \times n$ positive m-reciprocal matrix, then $I_{mEV} \geq 0$. Moreover, A is m-consistent if and only if $I_{mEV}(A) = 0$.*

To provide an m-inconsistency measure independently of the dimension n of the matrix A, Saaty in [37] proposed the consistency ratio CR_{mEV}. In order to distinguish it here from the other inconsistency measures, we shall call it the *m-EV-consistency ratio*. This is obtained by taking the consistency index I_{mEV} to its mean value R_{mEV}, i.e., the mean value of $I_{mEV}(A)$ of positive m-reciprocal matrices of dimension n, whose entries are uniformly distributed random variables on the interval $[1/9, 9]$, i.e.

$$CR_{mEV} = \frac{I_{mEV}}{R_{mEV}} . \tag{3.81}$$

The values of R_{mEV} can be found in the table in Sect. 2.5.1.

Example 3.6 Consider a PCM with all ones except for two corners (called a *corner comparisons matrix*, or *CPC*). The PC matrix $CPC(x, n)$, with $x > 1$, is defined by

$$CPC(x, n) = \begin{pmatrix} 1 & 1 & ... & 1 & x \\ 1 & 1 & ... & 1 & 1 \\ ... & ... & ... & ... & ... \\ 1 & 1 & ... & 1 & 1 \\ \frac{1}{x} & 1 & ... & 1 & 1 \end{pmatrix} . \tag{3.82}$$

It is not difficult to prove the following inequality (see [23]):

$$CI(CPC(x, n)) = \frac{\rho(CPC(x, n)) - n}{n - 1} \leq \frac{x}{n^2} \tag{3.83}$$

where $\rho(CPC(x, n)) = \lambda_{max}$ is a Perron eigenvalue of $CPC(x, n)$. In [35], Saaty proposed a heuristic threshold of 0.1 (i.e., 10%) for an inconsistent matrix to be considered as sufficiently consistent. This idea, however, proved to be unreasonable, as it is demonstrated below.

The inequality (3.83) has a very important implication. No matter how large x is, there is always a sufficiently large n that the left-hand side of (3.83) is as small as it can be assumed. Evidently, the arbitrary x in the matrix $CPC(x, n)$ of size n invalidates the acceptability of this matrix. Hence, the soundness of the eigenvalue-based inconsistency indicator represented by the left-hand side inequality (3.83) should be dismissed.

The $CPC(x, n)$ matrix in the above example shows that for the eigenvalue-based consistency index CI an error of an arbitrary value is acceptable for a large enough n (the matrix size). According to AHP theory, the $CPC(x, n)$ matrix is considered "consistent enough" for $CI(CPC(x, n)) \leq 0.1$, although it has two arbitrarily "erroneous elements" (x and $\frac{1}{x}$) in it.

In order to overcome the weaknesses of consistency index CI, Koczkodaj in [22] proposed his well-known consistency index KII, see Sect. 2.6, as

$$KII(A) = \max \left\{ 1 - \min \left\{ \frac{a_{ij}}{a_{ik}a_{kj}}, \frac{a_{ik}a_{kj}}{a_{ij}} \right\} | i, j, k \in \{1, \cdots, n\} \right\}. \tag{3.84}$$

where $A = \{a_{ij}\}$ is a PC matrix.

As is evident from (3.84), the value of Koczkodaj's consistency index belongs to the interval $[0; 1[$; PCM A is m-consistent if and only if $KII(A) = 0$, otherwise, A is inconsistent and $KII(A) > 0$. The more inconsistent A is, i.e., the greater is $\max \left\{ \frac{a_{ij}}{a_{ik}a_{kj}}, \frac{a_{ik}a_{kj}}{a_{ij}} \right\}$, the greater is index $KII(A)$.

In the following section, we shall deal with measuring the inconsistency of PCMs in a more general situation of alo-groups.

3.5.2 Measuring the Inconsistency of PCMs on Alo-Groups

Let $\mathscr{G} = (G, \odot, \leq)$ be a divisible alo-group and $A = \{a_{ij}\} \in PC_n(\mathscr{G})$. By Definition 3.2, A is $\odot$-inconsistent if at least one triple $\{c_i, c_j, c_k\}$ is $\odot$-inconsistent. The closeness to the consistency depends on the degree of consistency with respect to each 3-element subset $\{c_i, c_j, c_k\}$ and it can be measured by an *average* of these degrees. Let

$$e(i, j, k) = a_{ij} \odot a_{jk} \odot a_{ki}, \tag{3.85}$$

By Definition 1.16 of G-norm, we obtain

$$\|e(i, j, k)\| = \max\{e(i, j, k), (e(i, j, k))^{(-1)}\}. \tag{3.86}$$

Now, we define the *generalized average inconsistency index* $(GAI_\odot)$ *of the PC matrix* A as

$$GAI_\odot(A) = \left(\bigodot_{(i,j,k)\in T(A)} \|e(i, j, k)\| \right)^{(\frac{1}{n_T})}, \tag{3.87}$$

where by $T(A)$ we denote the set of 3-element subsets $\{a_{ij}, a_{jk}, a_{ik}\}$ of A with $i < j < k$. Clearly, $n_T = \frac{n!}{3!(n-3)!}$ is the cardinality of the set $T(A)$.

The Generalized Inconsistency index of the PC matrix A has already been defined in Definition 3.8 as

$$GI_\odot(A) = \max\{\|e(i, j, k)\| \, | i, j, k \in \{1, ..., n\}\}. \tag{3.88}$$

Remark 3.9 Inconsistency indexes, based on "averages", similarly to (3.87), suffer from the problem that the consistency condition has (by the definition) the requirement "for all triads" and with the growing PC matrix size (n), the value of inconsistency index "averages" all the values not taking into account possible local extreme values. If we accept it as normal, by the same logic, we should not worry about one nuclear bomb left behind since it is only one lost weapon for approximately 7 million inhabitants of this planet. Yet the "nuke" must be located and destroyed. The same goes for an inconsistency. That is why for measuring the inconsistency of a PCM the generalized inconsistency index $GI_\odot$ is clearly preferred to the generalized average inconsistency index $GAI_\odot$.

Last but not least, the $\odot$-inconsistency of PCM A can be measured also by the minimal error index $\mathscr{E}(A) = \mathscr{E}(A, w^*)$ defined by (3.36), where w^* is an optimal solution of problem (P), (3.33)–(3.35).

Example 3.7 Consider the matrix

$$A = \begin{pmatrix} 0.5\ 0.3\ 0.2 \\ 0.7\ 0.5\ 0.1 \\ 0.8\ 0.9\ 0.5 \end{pmatrix} \in PC_3(]0;\ 1[_m) \ .$$

Then by (3.85)–(3.88) we obtain

$$GAI_\odot(A) = GI_\odot(A) = \max\{ \frac{0.2 \cdot 0.7 \cdot 0.9}{0.2 \cdot 0.7 \cdot 0.9 + 0.3 \cdot 0.1 \cdot 0.8}, \frac{0.3 \cdot 0.8 \cdot 0.9}{0.2 \cdot 0.7 \cdot 0.9 + 0.3 \cdot 0.1 \cdot 0.8} \} =$$

$$= \max\{0.84, 0.16\} = 0.84 \ .$$

Example 3.8 ([9])

Consider the matrix

$$A = \begin{pmatrix} 1 & \frac{1}{7} & \frac{1}{7} & \frac{1}{5} \\ 7 & 1 & \frac{1}{2} & \frac{1}{3} \\ 7 & 2 & 1 & \frac{1}{9} \\ 5 & 3 & 9 & 1 \end{pmatrix} \in PC_4(\mathscr{R}_+) \ .$$

Then by (3.85), (3.86), and (3.87) we obtain

$$GAI_\odot(A) = \sqrt[4]{\|e(2,3,4)\| \cdot \|e(1,3,4)\| \cdot \|e(1,2,4)\| \cdot \|e(1,2,3)\|)} =$$

$$= \sqrt[4]{6 \cdot 12.6 \cdot 4.2 \cdot 2} = 5.02 \ .$$

By (3.85), (3.86), and (3.88) we obtain

$$GI_\odot(A) = \max\{e(2,3,4), e(1,3,4), e(1,2,4), e(1,2,3)\} =$$

$$= \max\{6.0, 12.6, 4.2, 2.0\} = 12.6 \ .$$

Here, in contrast to Example 3.7, $GAI_\odot(A) \neq GI_\odot(A)$.

3.6   Strong Transitive and Weak Consistent PCM

One of the most studied features of pairwise comparisons is their *coherence,* which is
a kind of compatibility of judgments. Several concepts for dealing with the coherence
of PCMs have been introduced and studied in the literature, such as (weak, strong)
consistency, (weak, strong) transitivity, see e.g., [9, 22, 33], and many others, and
the survey [24].

In Sect. 3.2 we have introduced the preference relation on the set of alternatives,
Here, we shall investigate more properties of transitive preference relations on the
finite set of alternatives $\mathscr{C}$ based on a pairwise comparisons matrix (PCM). Applying
the well-known concept of the transitivity of preference relations we introduce the
new concept of strong transitivity; it is weaker than the weak consistency introduced
in [11]. We show that for weak consistent preference relations and the corresponding
priority vector associated with the rows of the PCM by a strictly increasing aggregation function, we obtain the same preference order on the set of decision elements.
The strong transitivity is, however, stronger than the usual transitivity.

The following subsections deal with weak consistent and strong transitive PCMs.
Necessary and sufficient conditions for weak consistency derived in the literature,
[11], are mentioned, strong transitivity conditions are derived and discussed. At the
end, four examples characterizing the concepts and properties are presented. Finally,
conclusions close the section.

3.6.1   Special Notation

In this section we shall use the following special notation. Let $\mathscr{G} = (G, \odot, \leq)$
be a divisible alo-group and $A = \{a_{ij}\} \in PC_n(\mathscr{G})$. Let $v = (v_1, \cdots, v_n), w = (w_1, \cdots, w_n) \in G^n$. be two vectors in G^n.
We denote

$$v \stackrel{\leq}{=} w, \tag{3.89}$$

if $v_i \leq w_i \forall i \in \{1, \cdots, n\}$. In contrast to (3.89) we denote

$$v \leq w, \tag{3.90}$$

if $v_i \leq w_i \forall i \in \{1, \cdots, n\}$ and $\exists k \in \{1, \cdots, n\} : v_k < w_k$.
Moreover, we denote

$$v \lhd w, \tag{3.91}$$

if $v_i < w_i \forall i \in \{1, \cdots, n\}$.
Let $F : v \in G^n \to F(v) \in G$. If F is idempotent, symmetric and increasing, then
F is called an *aggregation function.*
Let us recall that

- F is *idempotent* if for any $a \in G$:

$$F(a, a, \cdots, a) = a; \tag{3.92}$$

- F is *symmetric* if for any vector $v = (v_1, v_2, \cdots, v_n) \in G^n$:

$$F(v_{i_1}, v_{i_2}, \cdots, v_{i_n}) = F(v_1, v_2, \cdots, v_n), \tag{3.93}$$

for every permutation $(v_{i_1}, v_{i_2}, \cdots, v_{i_n})$ of $\{1, \cdots, n\}$;
- F is *increasing* if

$$v_i \leq w_i \forall i \in \{1, \cdots, n\} \Rightarrow F(v_1, v_2, \cdots, v_n) \leq F(w_1, w_2, \cdots, w_n); \tag{3.94}$$

or, equivalently:

$$v \overset{\leq}{=} w \Rightarrow F(v) \leq F(w); \tag{3.95}$$

Moreover, F is *strictly increasing* if

$$v_i \leq w_i \forall i \in \{1, \cdots, n\} \text{ and } \exists k \in \{1, \cdots, n\} : v_k < w_k \Rightarrow \tag{3.96}$$

$$F(v_1, v_2, \cdots, v_n) < F(w_1, w_2, \cdots, w_n). \tag{3.97}$$

or, equivalently:

$$v \leq w \Rightarrow F(v) < F(w). \tag{3.98}$$

We also denote

- $a_i = (a_{i1}, a_{i2}, \cdots, a_{in})$—the ith row of $A = \{a_{ij}\}$,
- $a^j = (a_{1j}, a_{2j}, \cdots, a_{nj})$—the jth column of $A = \{a_{ij}\}$
- $n_0(a_i) = card(\{j | a_{ij} \geq e\}), A = \{a_{ij}\})$—the number of "nonnegative" elements of the ith row,
- $n_0(A) = (n_0(a_1), n_0(a_2), \cdots, n_0(a_n))$—n-vector of numbers $n_0(a_i)$.
- $m_\odot(v)$—mean of $v = (v_1, v_2, \cdots, v_n) \in G^n : m_\odot(v) = (\bigodot_{i=1}^n v_i)^{(\frac{1}{n})}$,
- $w_\odot(A) = (m_\odot(a_1), m_\odot(a_2), \cdots, m_\odot(a_n))$—n-vector of row means of A.

3.6.2 ⊙-Transitive PCM

The relation of *dominance* $\succ$ and relation of *indifference* $\simeq$ on the set $\mathscr{C}$, based on PCM $A = \{a_{ij}\}$ has already been defined in Sect. 3.2 as follows:

$$c_i \succ c_j \text{ if } a_{ij} > e, \tag{3.99}$$

$$c_i \simeq c_j \text{ if } a_{ij} = e. \tag{3.100}$$

If $c_i \succ c_j$, we say that c_i *strictly dominates* c_j, whereas if $c_i \simeq c_j$ we say that c_i and c_j are *indifferent*.

Combining the relation of strict dominance $\succ$ and indifference $\simeq$, we obtain the preference relation of *dominance* $\succeq$ on $\mathscr{C}$ as follows:

$$c_i \succeq c_j \text{ if } (c_i \succ c_j, \text{ or } c_i \simeq c_j), \text{ for all } c_i, c_j \in \mathscr{C}.$$

Transitivity is a basic property of coherence that the preference relation on alternatives should satisfy. We start with the definition of transitivity of PCMs generating transitivity on the set of alternatives $\mathscr{C}$.

Definition 3.9 A PCM $A = \{a_{ij}\}$ is $\odot$-*transitive*, if the following conditions hold:

$$a_{ij} \geq e \text{ and } a_{jk} \geq e \Rightarrow a_{ik} \geq e, \tag{3.101}$$

Definition 3.10 The preference relation of dominance $\succeq$ on $\mathscr{C} = \{c_1, \cdots, c_n\}$ is *transitive*, if for $c_i, c_j, c_k \in \mathscr{C}$:

$$c_i \succeq c_j \text{ and } c_j \succeq c_k \Rightarrow c_i \succeq c_k. \tag{3.102}$$

Proposition 3.10 *If PCM $A = \{a_{ij}\}$ is $\odot$-transitive, then the preference relation of dominance $\succeq$ on $\mathscr{C}$ defined by (3.99), (3.100) is transitive.*

Proposition 3.11 *If PCM $A = \{a_{ij}\}$ is $\odot$-consistent, then it is $\odot$-transitive.*

Proof Let $a_{ij} > e$ and $a_{jk} > e$, then by $\odot$-consistency, $a_{ik} = a_{ij} \odot a_{jk} > e \odot e = e$.
On the other hand, let $a_{ij} = e$ and $a_{jk} = e$, then $a_{ik} = a_{ij} \odot a_{jk} = e$. Hence, a is $\odot$-transitive. $\qquad\square$

Proposition 3.12 *PCM $A = \{a_{ij}\}$ is $\odot$-transitive if and only if there exists a vector $w = (w_1, \cdots, w_n)$ with $w_k \in G$ for all $k \in \{1, \cdots, n\}$, satisfying*

$$w_i > w_j \Leftrightarrow a_{ij} > e, \ w_i = w_j \Leftrightarrow a_{ij} = e. \tag{3.103}$$

Proof Let $A = \{a_{ij}\}$ be $\odot$-transitive PCM. By the main result from [10] and the notation in Sect. 3.5.1, $w_i = n_0(a_i), i \in \{1, \cdots, n\}$, satisfies (3.103).

Let $w = (w_1, w_2, \cdots, w_n)$ be a vector satisfying (3.103). Then the following implication holds:

$$(a_{ij} \geq e, a_{jk} \geq e) \Rightarrow (w_i \geq w_j, w_j \geq w_k) \Rightarrow w_i \geq w_k \Rightarrow a_{ik} \geq e,$$

thus, $A = \{a_{ij}\}$ is $\odot$-transitive as (3.103) is satisfied. $\qquad\square$

By (3.99), and (3.100), formula (3.103) is equivalent to

$$c_i \succ c_j \Leftrightarrow w_i > w_j, \ c_i \simeq c_j \Leftrightarrow w_i = w_j. \tag{3.104}$$

Hence, by Proposition 3.12, there exists an *ordinal ranking function on $\mathscr{C}$*

$$w : \mathscr{C} \to G,$$

such that for $x, y \in \mathscr{C}$:

$$x \succeq y \Leftrightarrow w(x) \geq w(y). \tag{3.105}$$

This means that we can reorder (or rename) the alternatives $c_i \in \mathscr{C}$ such that

$$c_1 \succeq c_2 \succeq ... \succeq c_n, \tag{3.106}$$

$$w_1 \geq w_2 \geq \cdots \geq w_n. \tag{3.107}$$

From now on we shall assume that this reordering of alternatives has been done. In other words, there exists a permutation $\pi : \{1, ..., n\} \to \{1, \cdots, n\}$, such that $A^* = \{a_{\pi(i),\pi(j)}\}$ is a PCM corresponding to the newly reordered set of alternatives satisfying (3.106).

Definition 3.11 Let $A = \{a_{ij}\}$ be an $\odot$-transitive PCM. A vector $w = (w_1, w_2, \cdots, w_n)$ with $w_k \in G$, for each $k \in \{1, \cdots, n\}$, satisfying

$$w_i > w_j \Leftrightarrow a_{ij} > e , w_i = w_j \Leftrightarrow a_{ij} = e \tag{3.108}$$

is a *coherent priority vector*.
   A vector $w = (w_1, w_2, \cdots, w_n)$ with $w_i \in G$, for each $i \in \{1, \cdots, n\}$, satisfying

$$w_i > w_j \Leftrightarrow a_{ij} < e , w_i = w_j \Leftrightarrow a_{ij} = e \tag{3.109}$$

is a *coherent anti-priority vector*.

3.6.3   Weak-$\odot$-Consistent PCM

In [11] a condition stronger than $\odot$-transitivity but weaker than $\odot$-consistency has been introduced, under which a class of vectors, including the vector of means of rows of $A = \{a_{ij}\}$, provides ordinal evaluation vectors for the actual ranking.

Definition 3.12 A PCM $A = \{a_{ij}\}$ is *weak-$\odot$-consistent*, if

$$a_{ij} \geq e, a_{jk} \geq e \Rightarrow a_{ik} = max\{a_{ij}, a_{jk}\}, \text{ if } a_{ij} = e, \text{ or } a_{jk} = e, \tag{3.110}$$

$$a_{ik} > max\{a_{ij}, a_{jk}\}, \text{ otherwise.} \tag{3.111}$$

The proof of the following proposition is straightforward and it can be found in [11].

Proposition 3.13 *The following assertions are equivalent:*

(i) $A = \{a_{ij}\}$ *is weak-$\odot$-consistent,*
(ii) $a_{ij} > e \Leftrightarrow a_i \rhd a_j$, $a_{ij} = e \Leftrightarrow a_i = a_j$, $\forall i, j \in \{1, \cdots, n\}$,
(iii) *each column a^j of A is a coherent priority vector, $j \in \{1, \cdots, n\}$,*
(iv) *each row a_i of A is a coherent anti-priority vector, $i \in \{1, \cdots, n\}$.*

The following proposition follows directly from Proposition 3.13 and the properties of the aggregation function (or see [11]).

Proposition 3.14 *If $A = \{a_{ij}\}$ is a weak $\odot$-consistent PCM and F is a strictly increasing the aggregation function, then $w_F(A) = (F(a_1), \cdots, F(a_n))$ is a coherent priority vector.*

A direct consequence of Proposition 3.14 is the following proposition. It says that the vector of GGMs of the rows of A is a coherent priority vector, i.e., PCM A satisfies the RP condition with respect to this priority vector.

Proposition 3.15 *Let $A = \{a_{ij}\}$ be a weak $\odot$-consistent PCM and let*

$$F(a_i) = \left(\bigodot_{j=1}^{n} a_{ij} \right)^{\left(\frac{1}{n}\right)}, i \in \{1, ..., n\}. \tag{3.112}$$

Then $F : G^n \to G$ is a strictly increasing aggregation function and the vector of row $\odot$-means of A, $m_\odot(A) = (m_\odot(a_1), ..., m_\odot(a_n))$, is a coherent priority vector.

Proof From the properties of group operation $\odot$ derived in Chap. 1, it follows that function $F(a_i)$ defined by (3.112) is a strictly increasing aggregation function. Hence, by Proposition 3.14, $m_\odot(A)$ is a coherent priority vector. $\square$

Example 3.9 Consider the set of four alternatives $\mathscr{C} = \{c_1, c_2, c_3, c_4\}$, and the corresponding PC matrix A given in the multiplicative alo-group $\mathscr{R}_+$ with $\odot = \cdot$, as follows:

$$A = \begin{pmatrix} 1 & 2 & 3 & 4 \\ \frac{1}{2} & 1 & 1 & 2 \\ \frac{1}{3} & 1 & 1 & 5 \\ \frac{1}{4} & \frac{1}{2} & \frac{1}{3} & 1 \end{pmatrix},$$

Then $n(A) = (3, 1, 1, 0)$ and $n_0(A) = (4, 3, 3, 1)$. By applying the geometric mean to each row of $A = \{a_{ij}\}$, we obtain

$$w_m(A) = (2.13, 1, 1.14, 0.4).$$

It is easy to check that A is $\cdot$-transitive and $n_0(A)$ is a coherent priority vector. Thus, by the chain

$$n_0(a_1) > n_0(a_2) = n_0(a_3) > n_0(a_4),$$

the ranking is $c_1 \succ c_2 \sim c_3 \succ c_4$.

Vector $w_m(A)$ is not a coherent priority vector, as $w_3 > w_2$, but $c_3 = c_2$.

3.6.4 Strong-$\odot$-Transitive PCM

Here, the main purpose is to establish a condition stronger than $\odot$-transitivity but weaker than *weak*-$\odot$-consistency, under which a class of vectors, including the vector of $\odot$-means of rows of A, provide coherent priority vectors. A relatively small difference in the definition of strong-$\odot$-transitivity in comparison to the definition of weak-$\odot$-consistency, where strict inequality relation "$<$" is substituted by "$\leq$" will have an essential impact.

Definition 3.13 A PCM $A = \{a_{ij}\}$ is *strong-$\odot$-transitive*, if

$$a_{ij} \geq e, a_{jk} \geq e \Rightarrow a_{ik} = max\{a_{ij}, a_{jk}\}, \text{ if } a_{ij} = e, \text{ or } a_{jk} = e, \quad (3.113)$$

$$a_{ik} \geq max\{a_{ij}, a_{jk}\}, \text{ otherwise.} \quad (3.114)$$

Remark 3.10 It follows from the definitions that if PCM A is weak-$\odot$-consistent, then A is strong-$\odot$-transitive. Moreover, if A is strong-$\odot$-transitive, then A is $\odot$-transitive (see Fig. 3.1).

Example 3.10 Consider the set of three alternatives $\mathscr{C} = \{c_1, c_2, c_3\}$, and the corresponding PC matrix A given as follows:

$$A = \begin{pmatrix} 1 & 2 & 3 \\ \frac{1}{2} & 1 & 3 \\ \frac{1}{3} & \frac{1}{3} & 1 \end{pmatrix}.$$

By Definition 3.13 it is clear that PCM A is strong-$\odot$-transitive, however, it is not weak-$\odot$-consistent.

In the following proposition we present the main results concerning the properties of PCMs.

Proposition 3.16 *The following assertions are equivalent:*

(i) $A = \{a_{ij}\}$ *is strong-$\odot$-transitive:*

(ii) $a_{ij} > e \Leftrightarrow a_i \geq a_j$, $a_{ij} = e \Leftrightarrow a_i = a_j$, $\forall i, j \in \{1, \cdots, n\}$,

(iii) $a_{ij} > e \Leftrightarrow a^i \leq a^j$, $a_{ij} = e \Leftrightarrow a^i = a^j$, $\forall i, j \in \{1, \cdots, n\}$.

(iv)   *each row a_i of A is a coherent anti-priority vector, $i \in \{1, \cdots, n\}$.*

Proof The proof is straightforward, even although, rather long. Thee full proof can be found in [21]. □

Remark 3.11 A direct consequence of Proposition 3.16 is that the vector of GGMs of the rows of A is a coherent priority vector, i.e., PCM A satisfies the RP (POP) condition with respect to this priority vector.

3.6.5 Examples

Example 3.11 Consider the set of four alternatives $\mathscr{C} = \{c_1, c_2, c_3, c_4\}$, and the corresponding PC matrices $C_j, j = 1, 2, 3, 4$, given in the multiplicative alo-group $\mathscr{R}_+$ with $\odot = \cdot$, as follows:

$$C_1 = \begin{pmatrix} 1 & 2 & 4 & 8 \\ \frac{1}{2} & 1 & 2 & 4 \\ \frac{1}{4} & \frac{1}{2} & 1 & 2 \\ \frac{1}{8} & \frac{1}{4} & \frac{1}{2} & 1 \end{pmatrix},$$

$$C_2 = \begin{pmatrix} 1 & 2 & 2 & 3 \\ \frac{1}{2} & 1 & 2 & 2 \\ \frac{1}{2} & \frac{1}{2} & 1 & 2 \\ \frac{1}{3} & \frac{1}{2} & \frac{1}{2} & 1 \end{pmatrix},$$

$$C_3 = \begin{pmatrix} 1 & 2 & 2 & 2 \\ \frac{1}{2} & 1 & 2 & 2 \\ \frac{1}{2} & \frac{1}{2} & 1 & 2 \\ \frac{1}{2} & \frac{1}{2} & \frac{1}{2} & 1 \end{pmatrix},$$

$$C_4 = \begin{pmatrix} 1 & 1 & 2 & 2 \\ 1 & 1 & 2 & 2 \\ \frac{1}{2} & \frac{1}{2} & 1 & 2 \\ \frac{1}{2} & \frac{1}{2} & \frac{1}{2} & 1 \end{pmatrix}.$$

As it can be easily verified, all four PCMs are $\cdot$-transitive. Moreover, C_1 is $\cdot$-consistent, C_2 weak-$\cdot$-consistent (and strong-$\cdot$-transitive), C_3 is strong-$\cdot$-transitive, but NOT weak-$\cdot$-consistent, and C_4 is neither strong-$\cdot$-transitive, nor weak-$\cdot$-consistent (see Fig. 3.1).

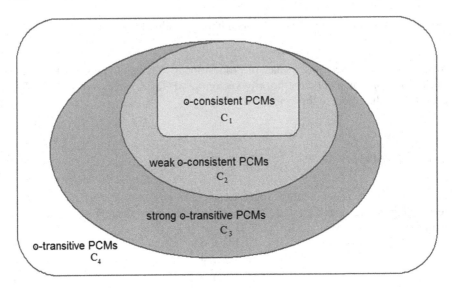

Fig. 3.1   Four PCMs

3.7   Pairwise Comparison Matrix with Missing Elements

3.7.1   Formulation of the Problem

In many decision-making processes, we assume that experts are capable of providing preference degrees between any pair of possible alternatives. However, this may not always be true, which creates a missing information problem. A missing value in the fuzzy pairwise comparison matrix $A = \{a_{ij}\}$ is not equivalent to a lack of preference for one alternative over another. A missing value can be the result of the incapacity of an expert to quantify the degree of preference of one alternative over another. In this case he/she may decide not to guess the degree of preference between some pairs of alternatives. It must be clear that when an expert is not able to express a particular value a_{ij}, because he/she does not have a clear idea of how the alternative c_i is better than alternative c_j, this does not mean that he/she prefers both options with the same intensity.

Sometimes, it is also pointed out that experts, when faced with comparisons between two alternatives, say c_i and c_j, would rather do not compare them directly. This could happen when, e.g., they do not yet have a good understanding of the preferences for this particular couple of alternatives. DMs may evade the answers, especially when taking a position on the given comparisons is morally or ethically uncomfortable, e.g., when comparing mortality risk vs. financial cost. Last but not least, some data may be lost or damaged. To model these situations, in the following we introduce the incomplete pairwise comparison matrix, or in other words, the PCM with missing elements.

The problem of missing elements in a PC matrix is well known and has been solved in different ways by many authors, e.g., Alonso et al. [1], Herrera-Viedma et al. [15, 20], Chiclana et al. [14], Fedrizzi and Giove [16], Kułakowski et al. [27], or Ramik [31].

The first approach (see [1, 14, 15]) is based on calculating the missing values of an incomplete additive PC matrix. This calculation is done by using only the known elements of the matrix, therefore by assuring that the reconstruction of the incomplete matrix is compatible with the rest of the information provided by the DM/expert. The primary purpose of this method is to maintain or maximize the global fa-consistency. The drawback is that it does not allow to deal with multiplicative PC matrices to be dealt with.

Let us now assume that one or more comparisons are missing. As a consequence, the PC matrix is incomplete and it is no longer possible to derive the priorities for the alternatives using the well-known methods of the EV, or GM, to mention only the most popular ones. Some methods have been proposed in the literature to solve the incomplete comparison problem. Most of these methods are formulated in the multiplicative framework [13, 18, 19, 29], some others in the additive framework [38, 39]. Let us very briefly describe the most important ideas presented in the above mentioned literature.

Two methods have been proposed by P. T. Harker. The first one [19], based on the GM method, is based on the concept of a "connecting path". If alternatives c_i and c_j are not compared with each other, let us denote by (c_i, c_j) the *missing comparison* (MC) and let c_{ij} be the corresponding numerical value to be estimated; a connecting path of size r has the following form:

$$c_{ij} = a_{ik_1}.a_{k_1 k_2} \cdots a_{k_r j} , \qquad (3.115)$$

where the comparison values at the right-hand side of (3.115) are known. The connecting path of size two, also called an elementary connecting path, corresponds to the more familiar expression

$$c_{ij} = a_{ik} a_{kj} . \qquad (3.116)$$

Note that each connecting path corresponds to an indirect comparison between c_i and c_j. Harker proposes that the value c_{ij} of the MC should be equal to the geometric mean of all connecting paths related to this MC. The drawback of this method is that the number of connecting paths grows with the number n of the alternatives in such a way that the method becomes computationally intractable for many real-world problems.

Harkers second method [19] is based on the following idea. The missing element (c_i, c_j) is set to be equal to w_i/w_j, where the components of the vector w are not known and are to be calculated. In other words, the missing entries are completed by setting them equal to the value they should approximate. The matrix obtained with the described substitution is denoted by C. An auxiliary nonnegative matrix A is then associated with C satisfying $Aw = Cw$. The matrix A is nonnegative and *quasi reciprocal*, in the sense that all its positive entries are reciprocal, but it contains

entries equal to zero. In this way, Harker transforms the original problem into that
of computing the principal eigenvector of a nonnegative quasi reciprocal matrix.
In order to justify his method, Harker develops a theory for such type of matrices,
following Saaty's method for positive reciprocal matrices.

In [38], Xu proposes to calculate the priority vector w of incomplete fuzzy preference matrix by a goal programming approach. This method minimizes the errors

$$\varepsilon_{ij} = |a_{ij} - 0.5 + 0.5(u_i - u_j)| \tag{3.117}$$

for all missing elements (x_i, x_j). He also proposes his goal programming approach
with another type of consistency.

In his second proposal, Xu in [39] develops a method, for an incomplete fuzzy
preference relations, similar to that introduced by Harker [18] for an incomplete
multiplicative PC matrix. In [39] the priority vector w is calculated by solving a
system of equations which corresponds to Harker's auxiliary eigenproblem.

Although the abovementioned authors have proposed methods for calculating the
ranking for incomplete pairwise comparisons, the influence of the incompleteness on
the final result has not been sufficiently studied. One of the exceptions here is Harker
[7, 18, 19]. These works, however, do not provide us with the methods for how to
measure incompleteness. Therefore, we study the impact of data incompleteness on
the correctness of ranking of the PCM, and in order to construct a tool for measuring
the level of incompleteness (see Ramik [31] and [27]).

As it was explained above, the first approaches are based on calculating the missing
values of an incomplete PCM. This calculation is done by using only the known
elements of the matrix for reconstruction of the incomplete matrix to be compatible
with the rest of the information provided by the DM. The primary purpose of these
methods is to maintain or maximize some global measure of consistency.

In this section we shall extend the abovementioned ideas to a more general situation of PCMs on alo-groups which enables us to solve the problem of PCMs for a
broader class of PCMs with missing elements.

3.7.2 Missing Elements of Matrix

In this section, $\mathscr{G} = (G, \odot, \leq)$ is assumed to be a divisible alo-group. For the sake
of simplicity of presentation, from now on we identify the alternatives $c_1, c_2, \cdots, c_n$
with integers $1, \cdots, n$, i.e., $\mathscr{C} = \{1, \cdots, n\}$ is the set of alternatives, $n > 1$. More-
over, let, $\mathscr{C} \times \mathscr{C} = \mathscr{C}^2$ be the Cartesian product of $\mathscr{C}$, i.e., $\mathscr{C}^2 = \{(i, j) | i, j \in \mathscr{C}\}$.
Let $K \subset \mathscr{C}^2$, $K \neq \mathscr{C}^2$ and $\mathscr{B}$ be the preference relation on K given by function
$\mu_{\mathscr{B}} : K \to S$, where $S \subset G$ is an appropriate scale. The preference relation $\mathscr{B}$ is
represented by the $n \times n$ *pairwise comparison matrix (PCM)* $B(K) = \{b_{ij}\}_K$ *with
missing elements denoted by the symbol* "?" depending on K as follows:

$$b_{ij} = \begin{cases} \mu_{\mathscr{B}}(i,j) \text{ if } (i,j) \in K \ , \\ ? \qquad \quad \text{if } (i,j) \notin K \ . \end{cases}$$

In what follows, the (i,j)th element of a PCM with missing elements will be denoted by symbol b_{ij} instead of symbol a_{ij} used in the preceding text.

We shall assume that each PC matrix $B(K) = \{b_{ij}\}_K$ with missing elements is $\odot$-reciprocal with respect to K, i.e.,

$$b_{ij} \odot b_{ji} = e \text{ for all } (i,j) \in K \ . \tag{3.118}$$

Let $L \subset K$, and $L = \{(i_1,j_1), (i_2,j_2), \cdots, (i_q,j_q)\}$ be a set of couples (i,j) of alternatives such that there exists a measured (evaluated) matrix element b_{ij}, with $b_{ij} \in G$ for all $(i,j) \in L$. The symmetric subset L' to L, which we denote by $L' = \{(j_1,i_1), (j_2,i_2), \cdots, (j_q,i_q)\}$, also belongs to K, i.e., $L' \subset K$. Hence, each subset K of $\mathscr{C}^2$ can be represented as follows:

$$K = L \cup L' \cup D,$$

where L is the set of couples of alternatives (i,j) given by the DM, L' is given by the reciprocity property, and D are indices of the diagonal of this matrix, i.e., $D = \{(1,1), (2,2), \cdots, (n,n)\}$, where $b_{ii} = e$ for all i. By the reciprocity property, if the DM is able to quantify b_{ij}, then he/she is able to quantify b_{ji}. The elements b_{ij} with $(i,j) \in \mathscr{C}^2$-K are called *the missing elements of matrix* $B(K)$. The missing elements of $B(K)$ are denoted by symbol "?" ("question mark"). On the other hand, the preference degrees given by the experts are denoted by b_{ij} where $(i,j) \in K$. By $\odot$-reciprocity it is sufficient that the expert will quantify only the elements b_{ij}, where $(i,j) \in L$, such that $K = L \cup L' \cup D$.

3.7.3 Problem of Missing Elements in PC Matrices Based on Optimization

Now, we shall deal with the problem of finding the values of missing elements of a given pairwise comparison matrix so that the extended matrix is as much $\odot$-consistent as possible. Ideally, the extended matrix will be $\odot$-consistent and we shall see that this happens if the induced graph to the extended matrix is a spanning tree (see the definition below).

Let $K \subset \mathscr{C}^2$, let $B(K) = \{b_{ij}\}_K$ be a pairwise comparison matrix with missing elements, let $K = L \cup L' \cup D$, where $L = \{(i_1,j_1), (i_2,j_2), \cdots, (i_q,j_q)\}$ is a set of couples of alternatives evaluated by the DM. Within this section we shall assume that

$$\mathscr{C} = \{1, \cdots, n\} \subset \{i_1, j_1, i_2, j_2, \cdots, i_q, j_q\} \ . \tag{3.119}$$

Remark 3.12 In the terminology of the *graph theory* condition (3.119) means that L is the *covering of* $\mathscr{C} = \{1, \cdots, n\}$.

Every PC matrix can also be naturally presented in the form of a graph (see e.g., [4]).

Definition 3.14 Let $\mathscr{C} = \{c_1, \cdots, c_n\}$ be a set of alternatives, $A = \{a_{ij}\}$ be a PCM with missing elements associated with $\mathscr{C}$. Let $\mathbf{G}(A) = (V, E)$ be a labeled, undirected graph with the set of vertices $V = \mathscr{C} = \{c_1, \cdots, c_n\}$ and the set of (undirected) edges $E = \{\{c_i, c_j\} | c_i, c_j \in V, c_i \neq c_j\}$ such that each $h \in E$ contains exactly two different elements, and if $h = \{c_i, c_j\}$, then a_{ij} is defined. $\mathbf{G}(A)$ is said to be *induced by matrix* A.

If for each $i, j \in \mathscr{C}$ the element a_{ij} of A is defined (i.e., there are no missing elements), then $\mathbf{G}(A)$ is said to be *complete*.

$\mathbf{G}(A)$ is the *tree*, if it is connected and it has no cyclic subgraphs, i.e.:

$$\{k_1, k_2\}, \{k_2, k_3\}, \cdots, \{k_{q-1}, k_q\}, \{k_q, k_1\}, q > 1.$$

$\mathbf{G}(A)$ is the *spanning tree*, if it is a tree and $V = \mathscr{C}$.

$\mathbf{G}(A) = (V, E)$ is the *covering*, if $\mathscr{C} \subset \{i_1, j_1, i_2, j_2, \cdots, i_q, j_q\}$, where

$$E = \{\{i_1, j_1\}, \{i_2, j_2\}, \cdots, \{i_q, j_q\}\}.$$

Remark 3.13 The number of edges denoted by $|E|$ of the spanning tree graph $\mathbf{G} = (V, E)$ is $|V| - 1$, see e.g., [4].

Definition 3.15 Let $K \subset \mathscr{C}^2$, let $B(K) = \{b_{ij}\}_K$ be a pairwise comparison matrix with missing elements, let $K = L \cup L' \cup D$. Let

$$\mathscr{E}(K, B(K), v) = \max_{(i,j)\in L} \{\|b_{ij} \odot v_j \div v_i\|\}, \qquad (3.120)$$

where $v_i, v_j \in G$ for all $i, j \in \{1, ..., n\}$.

The matrix $B^{\odot}(K) = \{b_{ij}^{\odot}\}_K$ called the $\odot$-*extension of* $B(K)$ is defined as follows:

$$b_{ij}^{\odot} = \begin{cases} b_{ij} & \text{if } (i,j) \in L, \\ b_{ji}^{(-1)} & \text{if } (i,j) \in L', \\ e & \text{if } (i,j) \in D, \\ v_i^* \div v_j^* & \text{if } (i,j) \notin K . \end{cases} \qquad (3.121)$$

Here, $v^* = (v_1^*, v_2^*, \cdots, v_n^*)$ called the $\odot$-*priority vector with respect to* K is the optimal solution of the following problem:

$(P_{\odot})$ $\mathscr{E}(K, B(K), v) \longrightarrow \min;$

subject to

$$\bigodot_{j=1}^{n} v_j = e, v_i \in G \text{ for all } i \in \{1, \cdots, n\}. \qquad (3.122)$$

Remark 3.14 In problem ($P_\odot$) we minimize the objective function only subject to the normalization condition (3.122) in order to obtain the $\odot$-priority vector with respect to K. We could, however, also ask to satisfy some more desirable properties: the POP, POIP, or RP conditions. Then we obtain a similar optimization problem to ($\odot$- Problem ε) in Sect. 3.3.2.

Remark 3.15 Notice, that the $\odot$-consistency index of the PC matrix $B^\odot(K) = \{b_{ij}^\odot\}_K$ is defined by (3.88) as $GI_\odot(B^\odot(K))$, or by (3.87) as $GAI_\odot(B^\odot(K))$. Moreover, if there are no missing elements, i.e., $K = L \cup L' \cup D = \mathscr{C}^2$, then $\mathscr{E}(K, B(K), v) = \mathscr{E}(B^\odot S(K), v)$, where $\mathscr{E}(B, v)$ is defined by (3.32).

Assumption (3.119) is natural as otherwise, if e.g., $k \in \{1, ..., n\}$ does not belong to $L = \{i_1, j_1, i_2, j_2, \cdots, i_q, j_q\}$, then v_k^* is not defined by problem ($P_\odot$) and therefore the $\odot$-extension of $B(K)$ cannot be defined by (3.121).

The following proposition states that problem ($P_\odot$) has always has an optimal solution on condition that (3.119) is satisfied. In the following subsections we show that for a special form of K, the optimal solution of ($P_\odot$) is unique and the $\odot$-extension of $B(K)$ is $\odot$-consistent. This result is important also from a more general point of view of the pairwise comparisons method and the generating of priority vectors. To the best of our knowledge, such result has not yet been published, at least in the context of the MCDM literature.

Proposition 3.17 *Let* $K = L \cup L' \cup D = \mathscr{C}^2$, *where* $L = \{(i_1, j_1), (i_2, j_2), \cdots, (i_{n-1}, j_{n-1})\}$ *such that*

$$\{1, \cdots, n\} \subset \{i_1, j_1, i_2, j_2, \cdots, i_{n-1}, j_{n-1}\}. \tag{3.123}$$

Then there exists a unique optimal solution $v^* = (v_1^*, v_2^*, \cdots, v_n^*)$ *of problem* ($P_\odot$) *such that* $B^\odot(K) = \{b_{ij}^\odot\}_K$ *is* $\odot$-consistent, *if and only if* $\mathscr{E}(K, B(K), v^*) = e$.

Proof The objective function of our optimization problem ($P_\odot$) is strictly quasi-concave, and therefore unimodal. Hence, the above mentioned situation that the iterative solution procedure can get "trapped" in a local minimum does not happen.

The rest of the proof follows from definitions (3.120) and (3.121). $\qquad\square$

Example 3.12 Consider alo-group $\mathscr{R} = (\mathbf{R}, +, \leq)$ from Example 3.1. Let $L = \{(1, 2), (1, 3), (2, 3), (3, 4)\}$, the expert evaluations be $b_{12} = 3, b_{23} = 5, b_{34} = 8$, with $b_{ij} + b_{ji} = 0$ for all $(i, j) \in L$, let $K = L \cup L' \cup D$. Hence $B(K) = \{b_{ij}\}_K$ is a PC matrix with missing elements as follows:

$$B(K) = \begin{pmatrix} 0 & 3 & 6 & ? \\ -3 & 0 & 5 & ? \\ ? & -5 & 0 & 8 \\ ? & ? & -8 & 0 \end{pmatrix}. \tag{3.124}$$

Solving (P_a) by the gradient method we obtain an optimal solution, i.e., a-priority vector u^* with respect to K, particularly, $u^* = (5.750, 3.417, -0.916, -8.251)$. By (3.121) we obtain $B^a(K)$-a-extension of $B(K)$ as follows:

$$B^a(K) = \begin{pmatrix} 0 & 3 & 6.667 & 14 \\ -3 & 0 & 5 & 11.667 \\ -6.667 & -5 & 0 & 8 \\ -14 & -11.667 & -8 & 0 \end{pmatrix}, \tag{3.125}$$

where $B^a(K)$ is a-inconsistent, as $\mathscr{E}(K, B(K), u^*) = 0.667$, hence $GAI_a(B^a(K)) = 0.667$.

Example 3.13 Consider alo-group $\mathscr{R} = (\mathbf{R}, +, \le)$ from Example 3.1. Let $L = \{(2, 3), (3, 4)\}$, the expert evaluations be $b_{12} = 3, b_{23} = 5, b_{34} = 8$, with $b_{ij} + b_{ji} = 0$ for all $(i, j) \in L$, let $K = L \cup L' \cup D$. Hence $B(K) = \{b_{ij}\}_K$ is a PC matrix with missing elements as follows:

$$B(K) = \begin{pmatrix} 0 & ? & ? & ? \\ -3 & 0 & 5 & ? \\ ? & -5 & 0 & 8 \\ ? & ? & -8 & 0 \end{pmatrix}. \tag{3.126}$$

Solving (P_a) by the gradient method (in Excell) we obtain a-priority vector u'^* with respect to K, specifically, $u'^* = (-8.290, 8.763, 3.763, -4.237)$. By (3.121) we obtain $B^a(K)$-a-extension of $B(K)$ as follows:

$$B^a(K) = \begin{pmatrix} 0 & -17.054 & -12.054 & -4.054 \\ 17.054 & 0 & 5 & 13 \\ 12.054 & -5 & 0 & 8 \\ 4.054 & -13 & -8 & 0 \end{pmatrix}, \tag{3.127}$$

where $B^a(K)$ is a-consistent, as $\mathscr{E}(K, B(K), u'^*) = 0$. Hence, $GAI_a(B^a(K)) = 0$.

Remark 3.16 The optimal solution in Example 3.13 is not unique.

3.7.4 Particular Cases of PC Matrices with Missing Elements

For a full definition of an $n \times n$ $\odot$-reciprocal pairwise comparison matrix we need $N = n(n - 1)/2$ pairs of elements to be evaluated by a DM/expert. For example, if $n = 10$, then $N = 45$, which is a considerable amount of pairwise comparisons. We ask that the DM to evaluate only $n - 1$ pairwise comparisons of alternatives in the form of a spanning tree. In this section we shall deal with two important particular cases of the spanning tree graph, namely, a path (sequence) and star. For the corresponding PCM with missing elements where the expert will evaluate only $n - 1$ pairwise comparisons of alternatives so that condition (3.119) is satisfied. Let $K \subset I^2$ be a set of indexes given by an expert, $B(K) = \{b_{ij}\}_K$ be a PCM with missing elements. Moreover, let $K = L \cup L' \cup D$.

First, we shall assume that the DM evaluates the matrix elements of L in the form of a spanning tree, in this case, a path (sequence). Without loss of generality we assume that $L = \{(i_1, j_1), (i_2, j_2), \cdots, (i_{n-1}, j_{n-1})\}$, particularly, the elements: $b_{12}, b_{23}, b_{34}, \cdots, b_{n-1,n}$.

3.7.5 Case $L = \{(1, 2), (2, 3), \cdots, (n - 1, n)\}$

Here, we assume that the expert evaluates $n - 1$ elements of the preference matrix $B(K)$, particularly, $b_{12}, b_{23}, b_{34}, \cdots, b_{n-1,n}$. First, we investigate the $\odot$-extension of $B^{\odot}(K)$. We derive the following result.

Proposition 3.18 Let $L = \{(1, 2), (2, 3), ..., (n - 1, n)\}$, $b_{ij} \in G$, $b_{ij} \odot b_{ji} = e$ for all $(i, j) \in K$, $K = L \cup L' \cup D$, and $L' = \{(2, 1), (3, 2), \cdots, (n, n - 1)\}$, $D = \{(1, 1), \cdots, (n, n)\}$. Then the optimal solution $v^* = (v_1^*, v_2^*, \cdots, v_n^*)$ of $(P_{\odot})$ is given as

$$v_1^* = \left(b_{12}^{(n-1)} \odot b_{23}^{(n-2)} \odot ... \odot b_{n-1,n}^{(1)}\right)^{\left(\frac{1}{n}\right)}, \tag{3.128}$$

$$v_k^* = b_{k-1,k}^{(-1)} \odot v_{k-1}, \text{ for all } k \in \{2, 3, \cdots, n\}. \tag{3.129}$$

Proof If (3.128) and (3.129) are satisfied, then

$$b_{i,i+1} = v_i^* \div v_{i+1}^* \text{ for } i \in \{1, \cdots, n - 1\},$$

and

$$v_1^* \odot v_2^* \odot \cdots \odot v_n^* = e,$$

which is (3.122). Hence,

$$\mathscr{E}(K, B(K), v^*) = \max_{(i,j) \in L} \{\|b_{ij} \odot v_j^* \div v_i^*\|\} = e, \tag{3.130}$$

the minimal value of the objective function subject to the given constraint (3.122). Therefore, $B^{\odot}(K) = \{v_i^{\div} * v_{i+1}^*\}$ is $\odot$-consistent. $\qquad \square$

Example 3.14 Consider alo-group $\mathscr{R} = (\mathbf{R}, +, \leq)$ from Example 3.1. Let $L = \{(1, 2), (2, 3), (3, 4)\}$, the expert evaluations be $b_{12} = 3, b_{23} = 5, b_{34} = 8$, with $b_{ij} + b_{ji} = 0$ for all $(i, j) \in L$, let $K = L \cup L' \cup D$. Hence $B(K) = \{b_{ij}\}_K$ is a PC matrix with missing elements as follows:

$$B(K) = \begin{pmatrix} 0 & 3 & ? & ? \\ -3 & 0 & 5 & ? \\ ? & -5 & 0 & 8 \\ ? & ? & -8 & 0 \end{pmatrix}. \tag{3.131}$$

Solving (P_a) we obtain a-priority vector v^* with respect to K, particularly, $v^* = (6.75, 3.75, -1.75, -9.25)$. By (3.121) we obtain a $B^a(K)$-a-extension of $B(K)$ as follows:

$$B^a(K) = \begin{pmatrix} 0 & 3 & 8 & 16 \\ -3 & 0 & 5 & 13 \\ -8 & -5 & 0 & 8 \\ -16 & -13 & -8 & 0 \end{pmatrix}, \tag{3.132}$$

where $B^a(K)$ is a-consistent, as $\mathscr{E}(K, B(K), v^*) = 0$, hence $GAI_a(B^a(K)) = 0$.

Example 3.15 Consider alo-group $\mathscr{R}_+ = (]0, +\infty[, \cdot, \le)$ from Example 3.2. Let $L = \{(1, 2), (2, 3), (3, 4)\}$, the expert evaluations be $b_{12} = 3, b_{23} = 5, b_{34} = 8$, with $b_{ij}.b_{ji} = 1$ for all $(i, j) \in L$, let $K = L \cup L' \cup D$. Hence $B(K) = \{b_{ij}\}_K$ is a PC matrix with missing elements as follows:

$$B(K) = \begin{pmatrix} 1 & 3 & ? & ? \\ \frac{1}{3} & 1 & 5 & ? \\ ? & \frac{1}{5} & 1 & 8 \\ ? & ? & \frac{1}{8} & 1 \end{pmatrix}. \tag{3.133}$$

Solving (P_m) we obtain m-priority vector w^* with respect to K, particularly, $w^* = (8.572, 2.857, 0.571, 0.071)$. Then we obtain $B^m(K)$ an m-extension of $B(K)$ as

$$B^m(K) = \begin{pmatrix} 1 & 3 & 15 & 120 \\ \frac{1}{3} & 1 & 5 & 40 \\ \frac{1}{15} & \frac{1}{5} & 1 & 8 \\ \frac{1}{120} & \frac{1}{40} & \frac{1}{8} & 1 \end{pmatrix}, \tag{3.134}$$

where, $B^m(K)$ is m-consistent, as $\mathscr{E}(K, B(K), w^*) = 1$, hence $GAI_m(B^m(K)) = 1$.

Secondly, we shall assume that the DM evaluates the matrix elements of L in the form of a spanning tree, in this case, a star. Without loss of generality we assume that $L = \{(i_1, j_1), (i_2, j_2), \cdots, (i_{n-1}, j_{n-1})\}$, particularly, the elements: $b_{12}, b_{13}, b_{14}, ..., b_{1,n}$.

3.7.6 Case $L = \{(1, 2), (1, 3), \cdots, (1, n)\}$

Now, we assume that the expert evaluates $n - 1$ elements of the preference matrix $B(K)$, particularly, $b_{12}, b_{13}, b_{14}, \cdots, b_{1,n}$. We derive the following result.

Proposition 3.19 *Let* $L = \{(1, 2), (1, 3), \cdots, (1, n)\}$, $b_{ij} \in G$, $b_{ij} \odot b_{ji} = e$ *for all* $(i, j) \in K$, $K = L \cup L' \cup D$, *and* $L' = \{(2, 1), (3, 1), \cdots, (n, 1)\}$, $D = \{(1, 1), ..., (n, n)\}$. *Then the optimal solution* $w^* = (w_1^*, w_2^*, \cdots, w_n^*)$ *of* $(P_\odot)$ *is given as*

$$w_1^* = \left(b_{12} \odot b_{23} \odot \cdots \odot b_{n-1,n}\right)^{\left(\frac{1}{n}\right)}, \tag{3.135}$$

$$w_k^* = b_{k-1,k}^{(-1)} \odot w_1, \text{ for all } k \in \{2, 3, \cdots, n\}. \tag{3.136}$$

Proof If (3.135) and (3.136) is satisfied, then

$$b_{1,i+1} = w_1^* \div w_{i+1}^* \text{ for } i \in \{1, \cdots, n-1\} ,$$

and

$$w_1^* \odot w_2^* \odot \cdots \odot w_n^* = e,$$

which is (3.122), hence

$$\mathcal{E}(K, B(K), w^*) = \max_{(i,j)\in L} \{\|b_{ij} \odot w_j^* \div w_i^*\|\} = e, \tag{3.137}$$

the minimal value of the objective function subject to the given constraint (3.122).

Therefore, $B^{\odot}(K) = \{w_i^{\div} * w_{i+1}^*\}$ is $\odot$-consistent. $\qquad\square$

Example 3.16 Consider alo-group $\mathcal{R} = (\mathbf{R}, +, \leq)$ from Example 3.1. Let $L = \{(1, 2), (1, 3), (1, 4)\}$, the expert evaluations be $b_{12} = 3, b_{13} = 5, b_{14} = 8$, with $b_{ij} + b_{ji} = 0$ for all $(i, j) \in L$, let $K = L \cup L' \cup D$. Hence $B(K) = \{b_{ij}\}_K$ is a PC matrix with missing elements as follows:

$$B(K) = \begin{pmatrix} 0 & 3 & 5 & 8 \\ -3 & 0 & ? & ? \\ -5 & ? & 0 & ? \\ -8 & ? & ? & 0 \end{pmatrix}. \tag{3.138}$$

Solving (P_a) we obtain a-priority vector v^* with respect to K, particularly, $v^* = (4, 1, -1, -4)$. By (3.121) we obtain a $B^a(K)$-a-extension of $B(K)$ as follows:

$$B^a(K) = \begin{pmatrix} 0 & 3 & 5 & 8 \\ -3 & 0 & 2 & 5 \\ -5 & -2 & 0 & 3 \\ -8 & -5 & -3 & 0 \end{pmatrix}, \tag{3.139}$$

where $B^a(K)$ is a-consistent, as $\mathcal{E}(K, B(K), v^*) = 0$, hence $GAI_a(B^a(K)) = 0$.

Example 3.17 Consider alo-group $\mathcal{R}_+ = (]0, +\infty[, \cdot, \leq)$ from Example 3.2. Let $L' = \{(1, 2), (2, 3), (3, 4)\}$, the expert evaluations be $b_{12} = 3, b_{23} = 5, b_{34} = 8$, with $b_{ij}.b_{ji} = 1$ for all $(i, j) \in L'$, let $K' = L' \cup L'' \cup D$. Hence $B(K') = \{b_{ij}\}_{K'}$ is a PC matrix with missing elements as follows:

$$B(K') = \begin{pmatrix} 1 & 3 & ? & ? \\ \frac{1}{3} & 1 & 5 & ? \\ ? & \frac{1}{5} & 1 & 8 \\ ? & ? & \frac{1}{8} & 1 \end{pmatrix}. \tag{3.140}$$

Solving (P_m) we obtain m-priority vector w^* with respect to K, particularly, $w^* = (8.572, 2.857, 0.571, 0.071)$. Then we obtain $B^m(K)$ an m-extension of $B(K)$ as

$$B^m(K') = \begin{pmatrix} 1 & 3 & 5 & 8 \\ \frac{1}{3} & 1 & \frac{5}{3} & \frac{8}{3} \\ \frac{1}{5} & \frac{3}{5} & 1 & \frac{8}{5} \\ \frac{1}{8} & \frac{3}{8} & \frac{5}{8} & 1 \end{pmatrix}, \tag{3.141}$$

where $B^m(K')$ is m-consistent as $\mathscr{E}(K', B(K'), w^*) = 1$, hence $GAI_m(B^m(K')) = 1$.

Several numerical examples based on fuzzy-additive and fuzzy-multiplicative alo-groups: $\mathscr{R}$ from Example 3.16 and $\mathscr{R}_+$ from Example 3.17, can be found in [17, 30, 31].

3.7.7 Incompleteness Index

Let $K \subset \mathscr{C}^2$, let $B = \{b_{ij}\}_K$ be a pairwise comparison matrix with missing elements, let $K = L \cup L' \cup D$, where $L = \{(i_1, j_1), (i_2, j_2), \cdots, (i_q, j_q)\}$ is a set of couples of alternatives evaluated by the DM. Each alternative i can be compared with, at most, $n - 1$ other alternatives. Therefore, the maximal value of $deg(c_i)$ for $i \in \{1, \cdots, n\}$ is $n - 1$. Here, the *degree of alternative* i, $deg(i)$, is defined by

$$deg(i) = \sharp(\{j | (i, j) \in \mathscr{C}^2\}), i \in \{1, \cdots, n\} \tag{3.142}$$

where symbol $\sharp(P)$ denotes the number of elements of P, see [27]. Similarly, the number of missing comparisons is given by $(n - 1) - deg(i)$. Hence, the expression

$$S(B) = \sum_{i=1}^{n} (n - 1 - deg(i))^2, \tag{3.143}$$

combines two features together: it rises when the number of missing comparisons increases and providing that there are two matrices of the same size and with the same number of missing comparisons, it is higher for a matrix that has larger irregularities in the distribution of missing elements. Then we define the *incompleteness index of matrix* B, $II(B)$, as

$$II(B) = \frac{SC(B)}{n(n-1)^2}. \tag{3.144}$$

which preserves both important features and its value is bounded and varies within the range $[0, (n-1)^2]$, see Kułakowski [27].

It is clear that $0 \leq II(B) \leq 1$. When the PC matrix is fully incomplete, i.e., there are no comparisons between alternatives, $II(B) = 1$. Conversely, if B is complete, i.e., all the alternatives are defined, $II(C) = 0$. Providing that the PCM is reciprocal, every alternative has to be compared with at least one different alternative.

Example 3.18 Consider the matrices $B(K)$ from Example 3.16 and $B(K')$ from Example 3.17. Let us calculate their incompleteness indices:

$$II(B(K)) = \frac{\sum_{i=1}^{4}(3 - deg(i))^2}{4.3^2} = \frac{0 + 4 + 4 + 4}{36} = 0.333. \qquad (3.145)$$

$$II(B(K')) = \frac{\sum_{i=1}^{4}(3 - deg(i))^2}{4.3^2} = \frac{4 + 1 + 1 + 4}{36} = 0.277 \qquad (3.146)$$

As we can see, the index of the first matrix is greater than the index of the second one, which reflects the fact that the distribution of the missing items in the rows of $B(K')$ is more aligned than in $B(K)$. However, both indices are quite small, as both matrices lack only 6 elements (out of 12).

3.8 Incompleteness—Conclusions

Here, we investigated a transitive preference relation on the finite set of decision elements (alternatives) based on a pairwise comparisons matrix (PCM). Considering the well-known concept of transitivity of preference relations we introduced the new concept of strong transitivity; it is weaker than the weak consistency introduced in the literature.

In this chapter we investigated PC matrices where relations among the entries are treated in four ways depending on a particular interpretation: multiplicatively, additively, fuzzy-multiplicatively, and fuzzy-additively. By various methods for deriving priorities from various types of preference matrices, we obtained the corresponding priority vectors for a final ranking of alternatives. Moreover, we derived some new results for situations where some elements of the fuzzy preference matrix are missing. Then, a unified framework for pairwise comparison matrices with missing elements based on alo-groups was presented. Two basic particular cases of PC matrices were discussed and illustrative numerical examples were presented. Finally, a tool for measuring incompleteness—the incompleteness index was proposed and discussed.

In the last part of this chapter, we are going to present a case study—an experiment—investigating a natural question: which alo-group or, in other words, which evaluation system among the most popular ones: additive, multiplicative, fuzzy, and Saaty's is the best from the viewpoint of DMs.

3.9 What Is the Best Evaluation Method for Pairwise Comparisons: A Case Study

3.9.1 Introduction to Case Study

In order to measure the preciseness of various types of evaluation methods of PCMs, we proposed an experiment; we used an opinion survey with a group of decision-makers (DMs): specifically, 187 students of the School of Business Administration, Silesian University in Opava, Czechia. The students received instructions on how to fill out the survey forms and completed the survey during their classes. On average, the experiment with students lasted half an hour. In the DM problem inspired by an example provided by Saaty in [36], participants were asked to make pairwise comparisons of areas of given geometrical figures presented in the survey form (see Fig. 3.2). Compared to the experiment described in [12], here the problem has a simpler structure, as there is no hierarchy there. The final goal is to rank the figures according to their area based on information given by their pairwise comparisons—data that were the main interest of the survey. The data are obtained by three different evaluation systems: multiplicative, additive, and fuzzy.

In contrast to the experiment in [12], here the exact result is known in advance, i.e., the values of weights of the priority vector are the exact areas calculated for individual geometric figures. Of course, the respondents (students) do not know them, and their task was to estimate the weights by the corresponding PCMs. Later on, calculating the weights from the given data, we obtained the deviations of the exact weights from the corresponding weights calculated from the estimates given by pairwise comparisons.

We start with some examples demonstrating the calculation of weights by the three evaluating systems.

3.9.2 Three Evaluation Systems

In the literature, three pairwise evaluating systems, special alo-groups of the real line **R**, are considered:

(1) If $a_{ij} \in [0; +\infty[$ represents a preference ratio, then $A = \{a_{ij}\}$ is a multiplicative PCM characterized by a *multiplicative evaluation system* (see Example 3.2).
(2) If $a_{ij} \in \mathbf{R} =]\infty; +\infty[$ represents a preference difference, then $A = \{a_{ij}\}$ is an additive PCM characterized by an *additive evaluation system* (see Example 3.1).
(3) If $a_{ij} \in [0; 1]$ reflects a fuzzy alo-group , then $A = \{a_{ij}\}$ is a fuzzy PCM characterized by a *fuzzy evaluation system* (see Example 3.3).

The foremost type of PCM, at least with respect to the number of real-world applications, is probably the multiplicative representation, used among others by Saaty in [36] in the theory of the Analytic Hierarchy Process (AHP) by using the finite semantic scale $\{\frac{1}{9}, \frac{1}{8}, \cdots, \frac{1}{2}, 1, 2, \cdots, 8, 9\}$ instead of evaluation system (1), i.e., characterized by the interval $]0; +\infty]$. This evaluation system is, essentially, a multiplicative system, but is not, however, an alo-group. We call it the *Saaty's evaluation system*.

Multiplicative, additive, fuzzy, and Saaty's evaluation systems have been already investigated in Chaps. 2 and 3, more detailed information is also presented below.

The result of the pairwise comparisons method based on a PCM, generally denoted as $G = \{g_{ij}\}$, is a ranking of set $\mathscr{C}$ of alternatives a mapping that assigns real values to the alternatives. Formally, it can be defined as the following function: The ranking function for $\mathscr{C}$ (the ranking of $\mathscr{C}$) is a function $w : \mathscr{C} \to \mathbf{R}_+$ that assigns to every alternative from $\mathscr{C} = \{c_1, \cdots, c_n\}$ a positive value from $\mathbf{R}_+$. Here, $w(x)$ represents the ranking value for $x \in \mathscr{C}$. The w function is usually written in the form of a vector of weights, i.e.,

$$w_G = (w_G(x_1), \cdots, w_G(x_n)) \tag{3.147}$$

and it is called the *G-priority vector* associated with the PCM G or, in short, the *priority vector*. Also, we say that priority vector w is generated by a priority generating method based on the PCM $G = \{g_{ij}\}$. A particular form of the ranking function is

$$w_G(x_i) = F(g_i), \quad i \in \{1, \cdots, n\}, \tag{3.148}$$

where $x_i \in \mathscr{C}$, $g_i = (g_{i1}, \cdots, g_{in})$ is a vector of elements of the ith row of $G = \{g_{ij}\}$ and $F(.)$ is an *aggregation function*, i.e., an *idempotent, symmetric and increasing function* (see [3]). A special aggregation function we shall deal with in this paper, is the $G - \odot$-mean of the PCMs row defined as follows:

$$w_G(x_i) = \mu_\odot(g_i) = \left(\bigodot_{j=1}^{n} g_{ij} \right)^{(\frac{1}{n})}, \quad i \in \{1, \cdots, n\}. \tag{3.149}$$

Then the priority vector associated with the PCM G we are interested in, is the *vector of G − ⊙-means*:

$$\mu_\odot(G) = (\mu_\odot(g_1), \cdots, \mu_\odot(g_n)). \tag{3.150}$$

Let $u, v \in G^n$, $u = (u_1, \cdots, u_n)$, and $v = (v_1, \cdots, v_n)$ be two vectors. The G-distance, $d_G(u, v)$, between vectors u and v is defined as $G - \odot$-mean:

$$d_g(u, v) = (d_G(u_1, v_1) \odot \cdots \odot d_G(u_n, v_n))^{(1/n)}. \tag{3.151}$$

In a similar way, we understand each PCM over alo-group $\mathscr{G} = (G, \odot, \leq)$ as a PCM $G = \{g_{ij}\}$ with entries in G. Here, general notions of reciprocity, consistency, and distance between PCMs are provided in this general algebraic context. Specifically, for additive, multiplicative, and fuzzy evaluations systems these notions are described as follows:

Additive Evaluation System

Let $G = \{g_{ij}\}$ be an additive PCM, then, $g_{ij} = 0$ if there is indifference between x_i and x_j, $g_{ij} > 0$ if and only if x_i is strictly preferred to x_j, whereas $g_{ij} < 0$ expresses the reverse preference. $G = \{g_{ij}\}$ is *additively reciprocal*, in short *a-reciprocal* if

$$g_{ij} \in G =]-\infty; +\infty[\text{ and } g_{ij} = -g_{ji} \ \forall i, j \in \{1, \cdots, n\}. \tag{3.152}$$

The aggregation function m_a we shall deal with, is the $G - \odot$-mean of the PCM's row, given by (3.149):

$$\mu_g(g_i) = \frac{1}{n} \sum_{j=1}^{n} g_{ij}, i \in \{1, \cdots, n\}. \tag{3.153}$$

The priority vector associated with the PCM G we are interested in is the vector of $G - \odot$-mean:

$$\mu_a(G) = (\mu_a(a_1), \cdots, \mu_a(a_n)). \tag{3.154}$$

The distance between two vectors $u, v \in G^n$, $u = (u_1, \cdots, u_n)$, $v = (v_1, \cdots, v_n)$ is as follows:

$$d_a(u, v) = \frac{1}{n} \sum_{i=1}^{n} |u_i - v_i|; \tag{3.155}$$

Evidently, $d_a(u, v)$ is greater than or equal to 0 and it is equal to 0 if and only if $u = v$.

Multiplicative Evaluation System

Let $M = \{m_{ij}\}$ be a multiplicative PCM, then, $m_{ij} = 1$ if there is indifference between x_i and x_j, $m_{ij} > 1$ if and only if x_i is strictly preferred to x_j, whereas $m_{ij} < 1$ expresses the reverse preference.

$M = \{m_{ij}\}$ is *multiplicatively reciprocal*, in short, *m-reciprocal* if

$$m_{ij} \in G =]0; +\infty[\text{ and } m_{ij} = 1/m_{ji} \ \forall i, j \in \{1, \cdots, n\}. \tag{3.156}$$

The aggregation function μ_m we shall deal with, is the $G - \odot$-mean of the PCM's row, given by (3.150):

$$\mu_m(m_i) = (\prod_{j=1}^{n} m_{ij})^{\frac{1}{n}}, i \in \{1, \cdots, n\}, \tag{3.157}$$

where $m_i = (m_{i1}, \cdots, m_{in})$ is a vector of elements of the ith row of $M = \{m_{ij}\}$. The priority vector associated with the PCM M we are interested in is the vector of $G - \odot$-means:

$$\mu_m(M) = (\mu_m(m_1), \cdots, \mu_m(m_n)). \tag{3.158}$$

The distance between two vectors $u, v \in G^n$, $u = (u_1, \cdots, u_n)$, $v = (v_1, \cdots, v_n)$ in the multiplicative evaluation system is the following one:

$$d_m(u, v) = \left(\prod_{i=1}^{n} \max\{\frac{u_i}{v_i}, \frac{v_i}{u_i}\} \right)^{\frac{1}{n}}. \tag{3.159}$$

Clearly, $d_m(u, v)$ is greater than or equal to 1 and it is equal to 1 if and only if $u = v$. Equation (3.159) is also used for computing the distance between two vectors obtained over the Saaty's scale.

Fuzzy Evaluation System

Let $F = \{f_{ij}\}$ be a fuzzy PCM, then, $f_{ij} = 0.5$ if there is indifference between x_i and x_j, $f_{ij} > 0.5$ if x_i is strictly preferred to x_j, whereas $f_{ij} < 0.5$ expresses the reverse preference.

$F = \{f_{ij}\}$ is *fuzzy reciprocal*, in short, *f-reciprocal* if

$$f_{ij} \in G =]0; 1[\text{ and } f_{ij} = 1 - f_{ji} \ \forall i, j \in \{1, \cdots, n\}. \tag{3.160}$$

The aggregation function μ_f we shall deal with in this paper is the $\mathscr{G}$-mean of the PCM's row is given by (3.150):

$$\mu_f(f_i) = \frac{(\prod_{j=1}^{n} f_{ij})^{\frac{1}{n}}}{(\prod_{j=1}^{n} f_{ij})^{\frac{1}{n}} + (\prod_{j=1}^{n}(1 - f_{ij}))^{\frac{1}{n}}}, i \in \{1, 2, ..., n\}, \tag{3.161}$$

where $f_i = (f_{i1}, \cdots, f_{in})$ is a vector of elements of the ith row of $F = \{f_{ij}\}$. The priority vector associated with the PCM F we are interested in is the vector of $G - \odot$-mean:

$$\mu_f(F) = (\mu_f(f_1), \cdots, \mu_f(f_n)). \tag{3.162}$$

The distance between two vectors $u, v \in G^n$, $u = (u_1, \cdots, u_n)$, $v = (v_1, \cdots, v_n)$ in the fuzzy evaluation system is as follows:

$$d_f(u, v) = \frac{\left(\prod_{i=1}^{n} d_f(u_i, v_i)\right)^{\frac{1}{n}}}{\left(\prod_{i=1}^{n} d_f(u_i, v_i)\right)^{\frac{1}{n}} + \left(\prod_{i=1}^{n}(1 - d_f(u_i, v_i))\right)^{\frac{1}{n}}}, \tag{3.163}$$

with

$$d_f(a, b) = \max\{\frac{a(1 - b)}{a(1 - b) + (1 - a)b}, \frac{(1 - a)b}{a(1 - b) + (1 - a)b}\}. \tag{3.164}$$

Here, $d_f(u, v)$ is greater than or equal to 0.5 and it is equal to 0.5 if and only if $u = v$.

Fig. 3.2 Geometric figures

3.9.3 The Experiment

In order to measure the preciseness of the estimates of various types of PCMs, we use an opinion survey with a sample of 187 students of master courses in economics and management at the Silesian University in Opava, School of Business Administration in Karvina, Czech Republic. The students completed the survey during a class period and received instructions on how to fill it out.

The topic of the survey was a DM problem inspired by an example provided by Saaty in [36]. The participants were asked to make pairwise comparisons of the size of areas of given geometrical figures (see Fig. 3.2).

In the design of the experiment, we decided that each participant should be allowed to complete only one survey. This "between-subject" designed experiment is more appropriate for our purpose than a "within-subject" designed experiment, where participants are exposed to more than one survey. It has long been observed that within-subject cannot be used when independence of multiple exposure is not guaranteed; the reader can refer to the literature mentioned, for example, in [12]. Thus, the respondents were divided into 4 groups, and the participants in each group had to fulfill exactly one of the questionnaires: Q^A, Q^M, Q^F, and Q^S.

The areas A_i, $i = 1, 2, 3, 4$ of the circle, rectangle, triangle, and square in Fig. 3.2 are

$$A_1 = 7.020, A_2 = 1.966, A_3 = 3.464, A_4 = 1 \tag{3.165}$$

respectively; thus, the exact additive, multiplicative and fuzzy PCMs are built directly from the known areas (3.165) as follows:

$$A^* = \{a_{ij}^*\} = \{A_i - A_j\} = \begin{pmatrix} 0.000 & 5.056 & 3.556 & 6.020 \\ -5.056 & 0.000 & -1.500 & 0.964 \\ -3.556 & 1.500 & 0.000 & 2.464 \\ -6.020 & -0.964 & -2.464 & 0.000 \end{pmatrix}, \tag{3.166}$$

$$M^* = \{m_{ij}^*\} = \{\frac{A_i}{A_j}\} = \begin{pmatrix} 1.000 & 3.575 & 2.027 & 7.020 \\ 0.280 & 1.000 & 0.567 & 1.964 \\ 0.493 & 1.764 & 1.000 & 3.464 \\ 0.142 & 0.509 & 0.289 & 1.000 \end{pmatrix}, \tag{3.167}$$

$$F^* = \{f_{ij}^*\} = \{\frac{A_i}{A_i + A_j}\} = \begin{pmatrix} 0.500 & 0.781 & 0.670 & 0.875 \\ 0.219 & 0.500 & 0.362 & 0.663 \\ 0.330 & 0.638 & 0.500 & 0.776 \\ 0.125 & 0.337 & 0.224 & 0.500 \end{pmatrix}. \qquad (3.168)$$

Let us consider the following isomorphisms (see [9, 12, 32, 34]) between multiplicative alo-group $\mathscr{R}_+ = (\mathbf{R}^+, \cdot, \leq)$ and additive alo-group $\mathscr{R} = (\mathbf{R}, +, \leq)$:

$$h : x \in \mathbf{R}^+ \to h(x) \in \mathbf{R} \quad h^{-1} : y \in \mathbf{R} \to h^{-1}(y) \in \mathbf{R}^+ \qquad (3.169)$$

Some examples are: $h(x) = \log_b x$ with the base $b > 1$, $h^{-1}(y) = b^y$ and isomorphism $q(x) = \frac{x}{1-x}$ between fuzzy alo-group $]0; 1[_m = (]0; 1[, \otimes, \leq)$ and multiplicative alo-group $\mathscr{R}_+ = (\mathbf{R}^+, \cdot, \leq)$ such that

$$q : x \in]0, 1[\to \frac{x}{1-x} \in \mathbf{R}^+, \quad q^{-1} : y \in \mathbf{R}^+ \to \frac{y}{1+y} \in]0; 1[. \qquad (3.170)$$

Remember that the goal of this experiment is to compute, by means of (3.155), (3.159), and (3.163), the distances between the exact PCMs (3.166), (3.167), and (3.168) and the PCMs provided by the students, in order to establish which approach (additive, multiplicative, fuzzy, and Saaty's) provides more preciseness. By the following propositions we transform the results from multiplicative and fuzzy evaluation systems to the same basis as the additive evaluation system, particularly through priority vectors, where the results are comparable. The proofs of the propositions are straightforward.

Proposition 3.20 *Let* $A^* = \{a_{ij}^*\} = \{A_i - A_j\}$, *where* $A_1, \ldots, A_n$ *are given values, and let for any additive PCM* $A^* = \{a_{ij}^*\}$, *and* $i, j \in \{1, \cdots, n\}$

$$w_A(x_i) = \frac{1}{n}\sum_{j=1}^{n}(a_{ij} - a_{i_0 j}) + 1, \qquad (3.171)$$

where i_0 *is an index of the benchmark alternative.*

Then (3.171) is a ranking function, $w_{\mathscr{C}} : \mathscr{C} \to] - \infty; +\infty[$, *such that for* $i \in \{1, \cdots, n\}$ *it holds that*

$$w_{A^*}(x_i) = \frac{1}{n}\sum_{j=1}^{n}(a_{ij}^* - a_{i_0 j}^*) + 1 = A_i. \qquad (3.172)$$

Remark 3.17 The *benchmark alternative*, i.e., index i_0, is the index of the alternative with (evidently) the smallest area considered as a unit, or more generally, the least "important" alternative considered as an "importance unit". The corresponding weight of this "benchmark" is set to 1 ("unity"), particularly in (3.165), where $i_0 = 4, A_4 = 1$.

Remark 3.18 By Proposition 3.20 we can "measure" the preciseness of evaluation of the additive PCM $A = \{a_{ij}\}$ calculating the distance $d_a(w_A, w_{A^*})$ between the priority vector (3.171) $w_A = (w_A(x_1), \cdots, w_A(x_n))$, associated with a PCM A and exact priority vector (3.165) $w_{A^*} = (A_1, \ldots, A_n)$, associated with the exact PCM $A^* = \{A_i - A_j\}$.

Proposition 3.21 *Let* $h :]0; +\infty[\rightarrow] - \infty; +\infty[$ *be an isomorphism between* $\mathscr{R}_+$ *and* $\mathscr{R}$, *and* $q :]0; 1[\rightarrow] - \infty; +\infty[$ *be an isomorphism between* $]0; 1[_m$ *and* $\mathscr{A}$. *Let* i_0 *be an index of the benchmark alternative.*
   (i) Let $M^* = \{m_{ij}^*\} = \{\frac{A_i}{A_j}\}$, *where* $A_1, \cdots, A_n$ *are given values, and let for any PCM* $M = \{m_{ij}\}$, *and* $i \in \{1, \cdots, n\}$ *define*

$$w_M(x_i) = h^{-1}\Big(\frac{1}{n} \sum_{j=1}^{n} h\Big(\frac{m_{ij}}{m_{i_0j}}\Big)\Big), \tag{3.173}$$

Then (3.173) is a ranking function, $w_M : \mathscr{C} \rightarrow] - \infty; +\infty[$, *such that for* $i \in \{1, 2, \cdots, n\}$ *it holds that*

$$w_{M^*}(x_i) = h^{-1}\Big(\frac{1}{n} \sum_{j=1}^{n} h\Big(\frac{m_{ij}^*}{m_{i_0j}^*}\Big)\Big) = A_i. \tag{3.174}$$

(ii) Let $F^* = \{f_{ij}^*\} = \{\frac{A_i}{A_i + A_j}\}$, *where* $A_1, \cdots, A_n$ *are given values, and let for any PCM* $F = \{f_{ij}\}$, *and* $i \in \{1, \cdots, n\}$ *define*

$$w_F(x_i) = h^{-1}\Big(\frac{1}{n} \sum_{j=1}^{n} h\Big(\frac{q(f_{ij})}{q(f_{i_0j})}\Big)\Big). \tag{3.175}$$

Then (3.175) is a ranking function, $w_F : \mathscr{C} \rightarrow] - \infty; +\infty[$, *such that for* $i \in \{1, \cdots, n\}$ *it holds that*

$$w_{F^*}(x_i) = h^{-1}\Big(\frac{1}{n} \sum_{j=1}^{n} h\Big(\frac{q(f_{ij}^*)}{q(f_{i_0j}^*)}\Big)\Big) = A_i. \tag{3.176}$$

Remark 3.19 Applying Proposition 3.21 we can measure the preciseness of the evaluation of multiplicative PCM $M = \{m_{ij}\}$ by calculating the distance $d_m(w_M, w_{M^*})$ and fuzzy PCM $F = \{f_{ij}\}$ by calculating the distance $d_f(w_F, w_{F^*})$ between the priority vector (3.175) $w_F = (w_F(x_1), \cdots, w_F(x_n))$, associated to a PCM F and exact priority vector $w_{F^*} = (A_1, \cdots, A_n)$, associated with the exact PCM $F^* = \{\frac{A_i}{A_i + A_j}\}$.

Example 3.19 Let us suppose that a student completes the multiplicative questionnaire Q^M and provides the following PCM:

Table 3.1 Means of imprecision P^X obtained by questionnaires Q^A, Q^M, Q^F, Q^S, and ANOVA results

P^A	C	P^F	P^S	p	F
0.963	0.892	0.685	0.679	0.013	3.715

$$M = \begin{pmatrix} 1.000 & 4.000 & 3.500 & 6.000 \\ 0.250 & 1.000 & 0.500 & 2.000 \\ 0.286 & 2.000 & 1.000 & 3.000 \\ 0.167 & 0.500 & 0.333 & 1.000 \end{pmatrix}.$$

Hence, by applying (3.173), with $h(x) = \log_2 x$, we have that priority vector $w_M = (w_M(x_1), \cdots, w_M(x_4)) = (7.416, 1.732, 2.803, 1.000)$ is associated with PCM M and exact priority vector (3.165) $w_{M^*} = (7.020, 1.966, 3.464, 1.000)$. The distance $d_m(w_M, w_{M^*}) = 0.322$ represents the student's imprecision.

3.9.4 Results of the Experiment

The results of the experiment described in the previous section are shown in the following tables. Approximately 11% of questionnaires had to be removed from the survey due to mistakes and misunderstanding in their content, the rest—168 of valid questionnaires were accepted for further statistical analysis. Here are the numbers of valid questionnaires, (see also Table 3.1):

$$Q1 - 45 \text{ questionnaires}, Q2 - 49, Q3 - 32, Q4 - 42.$$

For a statistical analysis of the data, one-factor ANOVA has been applied in the statistical package IBM-SPSS v. 21. Analysis of variance is used to test the hypothesis that several means are equal. This statistical method serves for testing, whether a factor—the particular type of evaluation method (here A, M, F, or S)—will influence the mean value of the distance between the corresponding PCM and the exact PCM. This distance is considered a measure of the preciseness of the corresponding method (see Table 3.2).

It can be seen from Table 3.2 that the least precise is the additive case (A) with the highest mean value of the mean distances. These differences are proved to be significant with the ANOVA F-test; indeed, p-value is less than 0.05, see second table in Table 3.1. However, if we remove the additive case from Table 3.2, then the differences are no longer significant; indeed, the p-value is higher than the usual confidence level $\alpha = 0.05$ (see Table 3.3).

Table 3.2 Results—ANOVA—4 methods

Anova: One factor

Descriptives

Type	N	Sum	Mean	Variance
Additive	45	43,343	0,963	0,233
Multiplicative	49	43,725	0,892	0,281
Fuzzy	32	21,926	0,685	0,155
Saaty's	42	28,522	0,679	0,234

ANOVA

Variability	SS	df	MS	F	p-value	F crit
Between groups	2,594	3	0,865	3,715	0,013	2,660
Within groups	38,169	164	0,233			
Total	40,763	167				

Table 3.3 Results—ANOVA—3 methods

Anova: One factor

Descriptives

Type	N	Sum	Mean	Variance
Multiplicative	49	43,725	0,892	0,281
Fuzzy	32	21,926	0,685	0,155
Saaty's	42	28,522	0,679	0,234

ANOVA

Variability	SS	df	MS	F	p-value	F crit
Between groups	1,308	2	0,654	2,813	0,064	3,072
Within groups	27,908	120	0,233			
Total	29,216	122				

3.9.5   Discussion and Conclusions

In this section, an experiment is performed in order to compare, from a behavioral point of view, four different preference evaluation approaches proposed in the literature for dealing with PCMs; i.e., the multiplicative, additive, and fuzzy approaches. Moreover, the multiplicative approach is verified in another version proposed by Saaty in AHP utilizing the popular nine-point scale.

Multiplicative, additive, and fuzzy preferences share the same algebraic structure (see [9]), i.e., an alo-group, whereas Saaty's modification does not share the structure of an alo-group. Here, we use properties for comparing the preciseness of multiplicative, additive, and fuzzy PCMs, in a real decision-making problem. By means of an opinion survey, DMs have been asked to express their subjective preferences about a decision-making problem with four alternatives—the areas of geometric figures. The results of the experiment show that by expressing preference ratios (i.e. the multiplicative preferences), or degrees of preference (i.e. fuzzy preferences), the DMs are more precise in their subjective preferences than when they express their preferences in differences (i.e., additive preferences). The additive method is, therefore, to be avoided for a higher preciseness. This conclusion supports the parallel results published in [12], concerning, however, the consistency of PCMs. When comparing the mean distance between multiplicative and fuzzy PCMs by the independent two-sample t-test, we find that there is no statistically significant difference between the means.

On the other hand, the same test has proven that there is a statistically significant difference between the means of multiplicative and Saaty's multiplicative PCMs. The method utilizing Saaty's nine-point scale is shown to be more precise than the standard multiplicative method where the scale is unbounded.

It could be that the source of imprecision is that the notion of "preference difference" has a less understandable meaning with respect to "preference ratio" or "degree of preference". The experiment seems to support the idea that Saaty's method with the ordinal nine-point scale is the most precise of all.

The behavioral operations research applied in this section may be an interesting approach when human behavior needs to be examined and taken into account.

References

1. Alonso S, Chiclana F, Herrera F, Herrera-Viedma E, Alcala-Fdes J, Porcel C (2008) A consistency-based procedure to estimate missing pairwise preference values. Int J Intell Syst 23:155–175 (2008)
2. Bana e Costa AA, Vasnick JA (2008) A critical analysis of the eigenvalue method used to derive priorities in the AHP. Eur J Oper Res 187(3):1422–1428
3. Barzilai J (1997) Deriving weights from pairwise comparison matrices. J Oper Res Soc 48(12):1226–1232
4. Bollobas J (2002) Modern graph theory. Springer, Berlin
5. Bourbaki N (1990) Algebra II. Springer, Heidelberg

6. Boyd S, Vandenberghe L (2004) Convex optimization. Cambridge University Press, Cambridge
7. Carmone FJ Jr, Kara A, Zanakis SH (1997) A Monte Carlo investigation of incomplete pairwise comparison matrices in AHP. Eur J Oper Res 102(3):538–553
8. Cavallo B, Brunelli M (2018) A general unified framework for interval pairwise comparison matrices. Int J Approx Reas 93:178–198
9. Cavallo B, D'Apuzzo L (2009) A general unified framework for pairwise comparison matrices in multicriteria methods. Int J Intell Syst 24(4):377–398
10. Cavallo B, D'Apuzzo L (2012) Deriving weights from a pairwise comparison matrix over an alo-group. Soft Comput 16:353–366
11. Cavallo B, D'Apuzzo L (2016) Ensuring reliability of the weighting vector: weak consistent pairwise comparison matrices. Fuzzy Sets Syst 296:21–34
12. Cavallo B, Ishizaka A, Olivieri MG, Squillante M (2018) Comparing inconsistency of pairwise comparison matrices depending on entries. J Oper Res Soc
13. Chen Q, Triantaphillou E (2001) Estimating data for multi-criteria decision making problems: optimization techniques. In: Pardalos PM, Floudas C (eds), Encyclopedia of optimization, vol 2. Kluwer Academic Publishers, Boston (2001)
14. Chiclana F, Herrera F, Herrera-Viedma E, Alonso S (2007) Some induced ordered weighted averaging operators and their use for solving group decision-making problems based on fuzzy preference relations. Eur J Operat Res 182:383–399
15. Chiclana F, Herrera-Viedma E, Alonso S (2009) A note on two methods for estimating pairwise preference values. IEE Trans Syst Man Cybern 39(6):1628–1633
16. Fedrizzi M, Giove S (2007) Incomplete pairwise comparison and consistency optimization. Eur J Operat Res 183(1):303–313
17. Gavalec M, Ramik J, Zimmermann K (2014) Decision making and optimization - special matrices and their applications in economics and management. Springer International Publishing, Switzerland
18. Harker PT (1987) Incomplete pairwise comparisons in the analytic hierarchy process. Math Model 9(11):837–848
19. Harker PT (1987) Alternative modes of questioning in the analytic hierarchy process. Math Model 9(3):353–360
20. Herrera-Viedma E, Herrera F, Chiclana F, Luque M (2004) Some issues on consistency of fuzzy preference relations. Eur J Operat Res 154:98–109
21. Jandova V, Talasova J (2013) Weak consistency: a new approach to consistency in the Saaty's analytic hierarchy process. Acta Universitatis Palackianae Olomucensis. Facultas Rerum Naturalium Mathematica 52(2):71–83
22. Koczkodaj WW (1993) A new definition of consistency of pairwise comparisons. Math Comput Model 18:79–93
23. Koczkodaj WW, Szwarc R (2014) On axiomatization of inconsistency indicators for pairwise comparisons. Fundam Inform 132:485–500
24. Kou G, Ergu D, Lin AS, Chen Y (2016) Pairwise comparison matrix in multiple criteria decision making. Technol Econ Dev Econ 22(5):738–765
25. Kułakowski K (2015) A heuristic rating estimation algorithm for the pairwise comparisons method. Cent Eur J Oper Res 23(1):187–203
26. Kulakowski K, Mazurek J, Ramik J, Soltys M (2019) When is the condition of order preservation met? Eur J Oper Res 277:248–254
27. Kułakowski K, Towards quantification of incompleteness in the pairwise comparisons methods. Int J Approx Reason, to appear
28. Mazurek J, Ramik J, Some new properties of inconsistent pairwise comparisons matrices. Internat J Approx Reason, to appear
29. Nishizawa K (2004) Estimation of unknown comparisons in incomplete AHP and it's compensation. Report of the Research Institute of Industrial Technology, Nihon University, p 77
30. Ramik J (2016) Incomplete preference matrix on alo-group and its application to ranking of alternatives. Internat J Math Oper Res 9(4):412–422

31. Ramik J (2014) Incomplete fuzzy preference matrix and its application to ranking of alternatives. Internat J Intell Syst 28(8):787–806
32. Ramik J (2015) Pairwise comparison matrix with fuzzy elements on alo-group. Inf Sci 297:236–253
33. Ramik J (2018) Strong reciprocity and strong consistency in pairwise comparison matrix with fuzzy elements. Fuzzy Optim Decis Mak 17:337–355
34. Ramik J, Vlach M (2013) Measuring consistency and inconsistency of pair comparison systems. Kybernetika 49(3):465–486
35. Saaty TL (1991) Multicriteria decision making - the analytical hierarchy process, vol I. RWS Publications, Pittsburgh
36. Saaty TL (2008) Relative measurement and its generalization in decision making: why pairwise comparisons are central in mathematics for the measurement of intangible factors, the analytic hierarchy/network process. RAC SAM Rev R Acad Cien Serie A Mat 102(2):251–318
37. Saaty TL (1980) Analytic hierarchy process. McGraw-Hill, New York
38. Xu ZS (2004) Goal programming models for obtaining the priority vector of incomplete fuzzy preference relation. Int J Approx Reason 36:261–270
39. Xu ZS (2005) A procedure for decision making based on incomplete fuzzy preference relation. Fuzzy Optim Decis Mak 4:175–189

Chapter 4
Pairwise Comparisons Matrices with Fuzzy and Intuitionistic Fuzzy Elements in Decision-Making

4.1 Introduction

The problem we consider here is the same as in Chaps. 2 and 3: Consider again, $\mathscr{C} = \{c_1, c_2, \cdots, c_n\}$ as a finite set of alternatives ($n > 1$). The goal of the DM is to rank the alternatives from the best to the worst (or, vice versa, which is equivalent), using the information given by the decision-maker in the form of an $n \times n$ pairwise comparisons matrix (PCM). The ranking of the alternatives is determined by the priority vector of real numbers $w = (w_1, w_2, \cdots, w_n)$ which is calculated from the corresponding PCM. There exist various methods for calculating the vector of weights based on the DM problem, and particularly, on the pairwise comparisons matrix, see e.g., [14, 17].

Fuzzy sets as the elements of the pairwise comparisons matrix can be applied in the DM problem whenever the decision-maker is not sure about the degree of preference his/her evaluations of the pairs in question. Fuzzy elements are useful for capturing the uncertainty stemming from the subjectivity of human thinking and from the incompleteness of information that is an integral part of multi-criteria decision-making problems. Fuzzy elements may also be useful as aggregations of crisp pairwise comparisons of a group of decision-makers in the group DM problem (see e.g., [3, 47]). Decision-makers usually acknowledge fuzzy pairwise preference data as imprecise knowledge about regular preference information. The preference matrix with fuzzy elements is then seen as a tool for constraining an ill-known precise consistent comparison matrix. In the literature the problem is also known as the *fuzzy analytic hierarchy process*, in short, FAHP (see e.g., [15, 21, 23, 28, 44] or [18]). A recent extensive review of various relevant aspects of the pairwise comparisons matrices such as measurement scales, consistency index, cardinal and ordinal inconsistency, missing data estimation, and priority derivation methods can be found in the survey paper [17].

Recently, in [18], a fuzzy eigenvector method for obtaining fuzzy weights from pairwise comparisons matrices with fuzzy elements has been presented. This method

© The Editor(s) (if applicable) and The Author(s), under exclusive license to Springer Nature Switzerland AG 2020
J. Ramik, *Pairwise Comparisons Method*, Lecture Notes in Economics and Mathematical Systems 690, https://doi.org/10.1007/978-3-030-39891-0_4

improves the methods published earlier in [13, 15]. However, the output is a normalized *fuzzy vector* of weights, which is not, from the perspective of DM, directly applicable to the final decision, and/or ranking of alternatives. Another suitable defuzzification or ranking procedure should be performed prior to making a final decision concerning the alternatives. The eigenvector method itself, however, is criticized by many authors for some unpleasant properties, see e.g., [2, 41]. Moreover, the concept of the *fuzzy eigenvalue* of the matrix with fuzzy elements does not seem to be properly justified, and the same could be said of the concept of the *fuzzy eigenvector*. For these reasons, we do not follow this way of the eigenvector method for deriving the priority vector of the pairwise comparisons matrix with fuzzy elements and propose our own approach.

On the other hand, Wang in [39] proposed a new definition of consistency with the triangular fuzzy preference relation (TFPR)—an extension of pairwise comparisons matrix (PC matrix) with fuzzy elements. However, Wang's definition of consistency based on the well-known "transitivity concept", is too restrictive: a PCM could be consistent (in this sense) only if all its elements are crisp. In other words, a consistent TFPR cannot be a PCM, and in particular, its diagonal elements are not "crisp ones" as it is natural in the case of usual PCM. Moreover, Wang's approach is applicable only for a PCMs with triangular fuzzy elements, and cannot be easily extended to fuzzy interval entries. Therefore, Wang's concept of consistency is not appropriate for a PCM with fuzzy elements being fuzzy intervals.

In [30, 31], the author of this monograph presented a general approach of PCM with fuzzy number elements based on alo-groups unifying the previous approaches. In [31], the concept of strong consistency was introduced and some relationships with the (ordinary) consistency were derived. In comparison to [30, 31], here, we extend our approach from fuzzy number entries (where the core is a singleton) to fuzzy intervals (where the core is an interval). Fuzzy intervals are entries of the PCM called the fuzzy pairwise comparisons matrix (FPCM).

Condition of order preservation (COP) was originally formulated within the context of the eigenvalue method (EVM). A criticism by [2] was directed at EVM and AHP. However, it could be easily noticed, that neither COP nor the notion of consistency does not depend on any method generating the priority vector. Hence the question arises whether the relationship between COP and the inconsistency is of a general nature and if so, what does this relationship look like? This chapter shall answer this question.

In this chapter we generalize the concept of COP to FPC matrices defining α-weak COP, α-strong COP and α-mean COP. Then, we reconsider the generalized geometric mean method (GGMM) and show that in satisfying "fuzzy" COP, the criteria under a generalized GMM depend on the locally defined inconsistency. Similarly to [20] as a consistency indicator we consider the Koczkodaj's inconsistency index [16]. Furthermore, we solve the problem of measuring the inconsistency of FPC matrices by defining a corresponding index.

In the second part of the chapter we deal with intuitionistic FPCMs (IFPCMs). We solve the problem of measuring the inconsistency of an IFPC matrix by defining corresponding indexes. The first index, called the consistency grade, is the maximal

α of the α-cut, such that the corresponding IFPC matrix is α-$\odot$-consistent. Moreover, the inconsistency index of the IFPC matrix is defined for measuring the fuzziness of this matrix by the distance of the IFPC matrix to the closest crisp consistent matrix. Consequently, an IFPC matrix is either crisp and consistent, or, it is inconsistent with the consistency grade <1 and inconsistency index greater than the identity element e. Within this chapter, we discuss a number of numerical examples in order to illustrate the proposed concepts and properties.

4.2 Preliminaries

The fundamental concept of this paper, the *pairwise comparisons matrix with fuzzy elements*, is based on the concept of the *fuzzy set*, which is built on level sets (or, alpha-cuts). Therefore, it will be useful to describe the fuzzy sets as special nested families of subsets of a set, particularly, the families of level sets. Here, we remember the definitions from Sect. 1.1 , or from publications [31, 34].

Definition 4.1 A *fuzzy subset* of a nonempty set X (or a *fuzzy set* on X) is a family $\{S_\alpha\}_{\alpha \in [0;1]}$ of subsets of X such that $S_0 = X$, $S_\beta \subset S_\alpha$ whenever $0 \leq \alpha \leq \beta \leq 1$, and $S_\beta = \cap_{0 \leq \alpha < \beta} S_\alpha$ whenever $0 < \beta \leq 1$. The *membership function of S* is the function μ_S from X into the unit interval $[0;1]$ defined by $\mu_S(x) = \sup\{\alpha \mid x \in S_\alpha\}$. Given $\alpha \in]0;1]$, the set $[S]_\alpha = \{x \in X \mid \mu_S(x) \geq \alpha\}$ is called the *α-cut of fuzzy set S*.

Usually, a fuzzy set S is defined by the *membership function* μ_S from X into the unit interval $[0;1]$ and then the family $\{S_\alpha\}_{\alpha \in [0;1]}$ of subsets of X is defined by the concept of α-cut $[S]_\alpha$, i.e., level sets of function μ_S. Our approach is the opposite, and the fuzzy set is a special family of sets and the concept of the membership function is defined later. Hence, the fuzzy set is based on the concept of a *set*, and not on the concept of a *function*. This approach is more appropriate and elegant.

Remember, that for X, a nonempty subset of the n-dimensional Euclidean space $\mathbf{R}^n$, a fuzzy set S in X is called *closed, bounded, compact*, or *convex* if the α-cut $[S]_\alpha$ is a closed, bounded, compact or convex subset of X for every $\alpha \in]0;1]$, respectively.

We say that a fuzzy subset S of $\mathbf{R}^* = \mathbf{R} \cup \{-\infty\} \cup \{+\infty\}$ is a *fuzzy interval* whenever S is normal and its membership function μ_S satisfies the following condition: there exist $a, b, c, d \in \mathbf{R}^*$, $-\infty \leq a \leq b \leq c \leq d \leq +\infty$, such that

$$\begin{aligned}
&\mu_S(t) = 0 \quad \text{if } t < a \text{ or } t > d, \\
&\mu_S \text{ is strictly increasing and continuous on the interval } [a;b], \\
&\mu_S(t) = 1 \quad \text{if } b \leq t \leq c, \\
&\mu_S \text{ is strictly decreasing and continuous on the interval } [c;d].
\end{aligned} \tag{4.1}$$

A fuzzy interval S is *bounded* if $[a;d]$ is a compact interval. For $\alpha = 0$ we define the *zero-cut* of S as $[S]_0 = [a;d]$. The nonnegative number $F(S) = d - a$ is called the *indeterminacy* of the bounded fuzzy interval S. A bounded fuzzy interval S is the *triangular fuzzy number* if $b = c$. We denote it by $S = (a, b, d)$. If $b < c$, then S is called the *trapezoidal fuzzy number*. We denote it by $S = (a, b, c, d)$. Notice that each crisp number is also a bounded fuzzy interval $S = (a, b, c, d)$ with $a = b = c = d$.

Moreover, if $a = b$ and $c = d$, then we obtain a *crisp interval case,* where the membership function is a characteristic function of a crisp interval. Notice that each crisp number is also a bounded fuzzy interval $S = (a, b, c, d)$ with $a = b = c = d$. Hence, our approach covers also evaluations of pairs by interval values, (and crisp values) which is usually dealt with as a separate problem.

In order to unify various approaches and prepare a more flexible presentation, similarly to Chap. 3, we apply alo-groups. Recall that an abelian group, [5], is a set, G, together with an operation $\odot$ that combines any two elements $a, b \in G$ to form another element in G denoted by $a \odot b$, see [5]. The symbol $\odot$ is a general placeholder for a concretely given operation. $(G, \odot)$ satisfies the following requirements known as the abelian group axioms, particularly: commutativity, associativity, there exists an identity element $e \in G$ and for each element $a \in G$ there exists an element $a^{(-1)} \in G$ called the inverse element to a. An ordered triple $(G, \odot, \leq)$ is said to be an abelian linearly ordered group, alo-group for short, if $(G, \odot)$ is a group, $\leq$ is a linear order on G, and for all $a, b, c \in G$

$$a \leq b \text{ implies } a \odot c \leq b \odot c. \tag{4.2}$$

By definition, an alo-group $\mathscr{G}$ is a lattice ordered group. Hence, there exists $\max\{a, b\}$, for each pair $(a, b) \in G \times G$. Nevertheless, a nontrivial alo-group $\mathscr{G} = (G, \odot, \leq)$ has neither the greatest element nor the least element. Because of the associative property, the operation $\odot$ can be extended by induction to an n-ary operation. $\mathscr{G} = (G, \odot, \leq)$ is *divisible* if for each positive integer n and each $a \in G$ there exists the (n)-th root of a denoted by $a^{(1/n)}$, i.e., $\left(a^{(1/n)}\right)^{(n)} = a$.

The well-known examples of alo-groups can be found in Chap. 3, Sect. 3.1.2 (see also [11] or [31]). Here, we present only a short survey of them.

Example 4.1 *Additive alo-group* $\mathscr{R} = (\mathbf{R}, +, \leq)$ is a continuous alo-group with

$$e = 0, \ a^{(-1)} = -a.$$

Example 4.2 *Multiplicative alo-group* $\mathscr{R}_+ = (\mathbf{R}_+, \cdot, \leq)$ is a continuous alo-group with
$$e = 1, \ a^{(-1)} = a^{-1} = 1/a.$$

Here, by $\cdot$ we denote the usual operation of multiplication.

Example 4.3 *Fuzzy additive alo-group* $\mathscr{R}_a = (\mathbf{R}, +_f, \leq)$, see [31], is a continuous alo-group with

$$a +_f b = a + b - 0.5, \ e = 0.5, \ a^{(-1)} = 1 - a.$$

Example 4.4 *Fuzzy multiplicative alo-group* $\mathscr{R}_m = (]0; 1[, \bullet_f, \leq)$, see [11], is a continuous alo-group with

$$a \bullet_f b = \frac{ab}{ab + (1-a)(1-b)}, e = 0.5, a^{(-1)} = 1 - a.$$

For more details, see Examples in Sect. 3.1.2.

4.3 FPC Matrices, Reciprocity, and Consistency

Our general approach based on alo-groups is useful, as it unifies various important approaches known from the literature. This fact has already been demonstrated on the four examples presented above, where the well-known alo-groups are shown. In particular, all the concepts and properties that will be presented below can be easily applied to any alo-group. In practice, the type of applied alo-group applied depends on the particular DM problem. Sometimes, it is more appropriate to apply, e.g., an additive alo-group (Example 4.1) (see [40]), in other cases, the multiplicative alo-groups (Example 4.2) (see [47]), or fuzzy alo-groups (Examples 4.3 and 4.4, see [44]), are useful from the point of view of interpretation.

Before we shall investigate PC matrices with fuzzy elements we remember some concepts and properties of PC matrices on an alo-group with crisp elements.

A crisp PC matrix $A = \{a_{ij}\}$ is said to be $\odot$-*reciprocal*, if the following condition holds: For every $i, j \in \{1, \cdots, n\}$

$$a_{ij} \odot a_{ji} = e, \tag{4.3}$$

or, equivalently,

$$a_{ji} = a_{ij}^{(-1)}. \tag{4.4}$$

A crisp FPC matrix $A = \{a_{ij}\}$ is $\odot$-*consistent* if for all $i, j, k \in \{1, \cdots, n\}$

$$a_{ik} = a_{ij} \odot a_{jk}, \tag{4.5}$$

or, equivalently,

$$a_{ij} \odot a_{jk} \odot a_{ki} = e.$$

Remember that an $\odot$-consistent PC matrix $A = \{a_{ij}\}$ is $\odot$-reciprocal, but not vice versa. The following equivalent condition for consistency of PC matrices is well known, see Proposition 3.4 or e.g., [11, 35]. A crisp PC matrix $A = \{a_{ij}\}$ is $\odot$-consistent if and only if there exists a vector $w = (w_1, w_2, \cdots, w_n), w_i \in G$, such that

$$a_{ij} = w_i \div w_j \text{ for all } i, j \in \{1, \cdots, n\}. \tag{4.6}$$

Here, $w_i \div w_j = w_i \odot w_j^{(-1)}$.

Now, we extend the above stated definition of $\odot$-reciprocity and $\odot$-consistency to non-crisp matrices with fuzzy elements, (see also [31]). In particular, we introduce new concepts of reciprocity and consistency based on α-cuts: α-$\odot$-reciprocity and α-$\odot$-consistency. We start, however, with the α-$\odot$-reciprocity in the crisp case.

Let $\mathscr{G} = (G, \odot, \leq)$ be a divisible and continuous alo-group over an open interval G of $\mathbf{R}$. Let $\alpha \in [0; 1]$, $\tilde{A} = \{\tilde{a}_{ij}\}$ be an $n \times n$ matrix, where each element is a bounded fuzzy interval of the alo-group $\mathscr{G}$, and let

$$[\tilde{a}_{ij}]_\alpha = [a_{ij}^L(\alpha); a_{ij}^R(\alpha)], \tag{4.7}$$

be an α-cut of $\tilde{a}_{ij}$, $i, j \in \{1, \cdots, n\}$.

Matrix $\tilde{A} = \{\tilde{a}_{ij}\}$ is said to be α-$\odot$-*reciprocal*, if the following two conditions hold for each $i, j \in \{1, \cdots, n\}$:

$$a_{ii}^L(\alpha) = a_{ii}^R(\alpha) = e, \tag{4.8}$$

$$a_{ij}^L(\alpha) \odot a_{ji}^R(\alpha) = e. \tag{4.9}$$

If $\tilde{A} = \{\tilde{a}_{ij}\}$ is α-$\odot$-reciprocal for all $\alpha \in [0; 1]$, then it is called $\odot$-*reciprocal*.

If $\tilde{A} = \{\tilde{a}_{ij}\}$ is $\odot$-reciprocal, then $\tilde{A} = \{\tilde{a}_{ij}\}$ is called the *fuzzy pairwise comparisons matrix, fuzzy PC matrix, FPC matrix*, or, in short, *FPCM*.

Remark 4.1 Here, by (4.8) we assume that all diagonal elements of the FPC matrix are crisp, and, in particular, they are equal to the identity element e of $\mathscr{G}$. An interpretation of this assumption is natural: comparing an alternative with itself, the result of the evaluation is: *"equal"*, which is denoted by e.

Moreover, (4.9) is equivalent to

$$a_{ij}^R(\alpha) \odot a_{ji}^L(\alpha) = e. \tag{4.10}$$

In practice, e.g., in AHP (see e.g., [35]), the crisp PC matrix is assumed to be reciprocal. Similarly, here, we ask that each FPC matrix is $\odot$-reciprocal. In practice, this assumption is not too restrictive as in a FPCM $\tilde{A} = \{\tilde{a}_{ij}\}$, the decision-maker usually evaluates at most $N = \frac{n(n-1)}{2}$ elements $\tilde{a}_{ij}$ for $i, j \in \{1, \cdots n\}$, such that $c_i \succ c_j$. The elements $\tilde{a}_{ji}$ are automatically taken as the $\odot$-reciprocal ones to $\tilde{a}_{ij}$, i.e., satisfying (4.9), as well as (4.8). Evidently, if $\tilde{A}$ is crisp, then the definition is equivalent to Definition 4.3.

Remark 4.2 Notice that in practice, elements of pairwise comparisons matrices may be either crisp or fuzzy numbers, and/or fuzzy intervals, and/or fuzzy intervals with bell-shaped membership functions, triangular fuzzy numbers, trapezoidal fuzzy numbers, etc. On the other hand, the membership functions of the fuzzy elements are not necessarily piecewise linear, which is a usual requirement in the literature (see e.g.,[18, 28, 39]). In our approach, only monotonicity of the membership function is required in the definition of bounded fuzzy interval. Such fuzzy elements may either be evaluated by individual decision-makers or they may be made up of crisp pairwise evaluations of decision-makers in a group DM problem (see e.g., [3]). The rationality and compatibility of a decision-making process can be achieved by the consistency property which will be studied in the sequel.

Now, we turn to two concepts of consistency of FPC matrices. We start with the definition of *weak* α-$\odot$-consistent FPC matrix. Later on, we shall define a stronger concept, particularly, an α-$\odot$-consistency of FPC matrix.

Definition 4.2 Let $\alpha \in [0; 1]$. A FPC matrix $\tilde{A} = \{\tilde{a}_{ij}\}$ is said to be *weak* α-$\odot$-*consistent*, if the following condition holds
There exists a crisp matrix $A' = \{a'_{ij}\}$ with $a'_{ik} \in [\tilde{a}_{ik}]_\alpha, a'_{ij} \in [\tilde{a}_{ij}]_\alpha, a'_{jk} \in [\tilde{a}_{jk}]_\alpha$, such that $A' = \{a'_{ij}\}$ is consistent, i.e., for each $i, j, k \in \{1, \cdots, n\}$ it holds that

$$a'_{ik} = a'_{ij} \odot a'_{jk}. \tag{4.11}$$

The FPC matrix $\tilde{A} = \{\tilde{a}_{ij}\}$ is said to be *weak* $\odot$-*consistent*, if $\tilde{A}$ is weak α-$\odot$-consistent for all $\alpha \in [0; 1]$.
If for some $\alpha \in [0; 1]$ the FPC matrix $\tilde{A} = \{\tilde{a}_{ij}\}$ is not weak α-$\odot$-consistent, then $\tilde{A}$ is called α-$\odot$-*inconsistent*.

Remark 4.3 In other words, Definition 4.2 says that an FPC matrix $\tilde{A} = \{\tilde{a}_{ij}\}$ is weak $\odot$-consistent if there exists a crisp *consistent* matrix $A' = \{a'_{ij}\}$, contained in the α-cuts of its elements, i.e., $a'_{ik} \in [\tilde{a}_{ik}]_\alpha, a'_{ij} \in [\tilde{a}_{ij}]_\alpha, a'_{jk} \in [\tilde{a}_{jk}]_\alpha$ for each $i, j, k \in \{1, \cdots, n\}$.

Remark 4.4 Let $\alpha, \beta \in [0; 1], \alpha \geq \beta$. Evidently, if $\tilde{A} = \{\tilde{a}_{ij}\}$ is weak α-$\odot$-consistent, then it is weak β-$\odot$-consistent.

Remark 4.5 It can be easily verified, that (4.9) holds for all $i, j \in \{1, \cdots, n\}$ if and only if (4.9) holds for all $i, j \in \{1, \cdots, n\}, 1 \leq i < j \leq n$.
Similarly, (4.11) holds for all $i, j, k \in \{1, \cdots, n\}$ if and only if (4.11) holds for all $i, j, k \in \{1, \cdots, n\}, 1 \leq i < j < k \leq n$.

The next proposition gives three necessary and sufficient conditions for an FPC matrix to be weak α-$\odot$-consistent. Condition (ii) is in a sense an extension of Proposition 3.4.

Proposition 4.1 *Let* $\alpha \in [0; 1]$, *let* $\tilde{A} = \{\tilde{a}_{ij}\}$ *be a FPC matrix,* $[\tilde{a}_{ij}]_\alpha = [a^L_{ij}(\alpha); a^R_{ij}(\alpha)]$ *be an* α-*cut of* $\tilde{a}_{ij}$. *The following conditions are equivalent.*

(i) $\tilde{A} = \{\tilde{a}_{ij}\}$ *is weak* α-$\odot$-*consistent.*
(ii)   *There exists a vector* $w = (w_1, \cdots, w_n)$ *with* $w_i \in G, i \in \{1, \cdots, n\}$, *such that for each* $i, k \in \{1, \cdots, n\}$, *it holds*

$$a^L_{ik}(\alpha) \leq w_i \div w_k \leq a^R_{ik}(\alpha). \tag{4.12}$$

(iii)   *For each $i, j, k \in \{1, \cdots, n\}$, it holds that*

$$a_{ik}^L(\alpha) \leq a_{ij}^R(\alpha) \odot a_{jk}^R(\alpha), \qquad (4.13)$$

(iv)   *For each $i, j, k \in \{1, \cdots, n\}$, it holds*

$$a_{ik}^R(\alpha) \geq a_{ij}^L(\alpha) \odot a_{jk}^L(\alpha). \qquad (4.14)$$

Proof Let $\alpha \in [0; 1]$.
$(i) \Rightarrow (ii)$:
By Definition 4.2, there exists a crisp consistent matrix $A' = \{a'_{ij}\}$ such that for each $i, k \in \{1, \cdots, n\}$ it holds that

$$a_{ik}^L(\alpha) \leq a'_{ik} \leq a_{ik}^R(\alpha). \qquad (4.15)$$

Let $j_0 \in \{1, \cdots, n\}$ be arbitrary and fixed, then by consistency of A' we have

$$a'_{ik} = a'_{ij_0} \odot a'_{j_0 k}.$$

Setting $w_i = a'_{ij_0}$ for $i \in \{1, \cdots, n\}$, we obtain for each $i, k \in \{1, \cdots, n\}$ from (4.15)

$$a_{ik}^L(\alpha) \leq w_i \div w_k \leq a_{ik}^R(\alpha). \qquad (4.16)$$

$(ii) \Rightarrow (i)$:
Let $w = (w_1, \cdots, w_n)$ be a vector satisfying (4.12) for each $i, k \in \{1, \cdots, n\}$. Setting $a'_{ik} = w_i \div w_k$ for each $i, k \in \{1, \cdots, n\}$, we obtain a crisp $\odot$-consistent matrix $A' = \{a'_{ij}\}$ such that for each $i, j, k \in \{1, \cdots, n\}$

$$a_{ik}^L(\alpha) \leq a'_{ik} \leq a_{ik}^R(\alpha),$$

and $a'_{ik} = a'_{ij} \odot a'_{jk}$, hence, $\tilde{A} = \{\tilde{a}_{ij}\}$ is weak α-$\odot$-consistent.
$(i) \Rightarrow (iii)$:
By (i) we have a crisp $\odot$-consistent matrix $A' = \{a'_{ij}\}$ such that for each $i, j, k \in \{1, \cdots, n\}$ it holds $a'_{ij} \leq a_{ij}^R(\alpha)$, $a'_{jk} \leq a_{jk}^R(\alpha)$, hence by "multiplying" the left sides as well as right sides, we obtain

$$a'_{ij} \odot a'_{jk} \leq a_{ij}^R(\alpha) \odot a_{jk}^R(\alpha),$$

and by consistency of $A' = \{a'_{ij}\}$

$$a'_{ik} \leq a_{ij}^R(\alpha) \odot a_{jk}^R(\alpha),$$

for each $i, j, k \in \{1, \cdots, n\}$. As $a_{ik}^L(\alpha) \leq a'_{ik}$, we obtain

$$a_{ik}^L(\alpha) \le a_{ik}' \le a_{ij}^R(\alpha) \odot a_{jk}^R(\alpha),$$

i.e., (4.13) is satisfied.

$(iii) \Rightarrow (iv)$:

Let for all $i, j, k \in \{1, \cdots, n\}$, (4.13) it holds that

$$a_{ik}^L(\alpha) \le a_{ij}^R(\alpha) \odot a_{jk}^R(\alpha).$$

Then by applying α-$\odot$-reciprocal condition, i.e., for all $r, s \in \{1, \cdots, n\}$ it holds that

$$a_{rs}^L(\alpha) = (a_{sr}^R(\alpha))^{(-1)},$$

and we obtain

$$(a_{ki}^R(\alpha))^{(-1)} \le (a_{kj}^L(\alpha) \odot a_{ji}^L(\alpha))^{(-1)},$$

or, equivalently

$$a_{ki}^R(\alpha) \ge a_{kj}^L(\alpha) \odot a_{ji}^L(\alpha).$$

The last inequality holds for all $i, j, k \in \{1, \cdots, n\}$. Hence, (4.14) is satisfied.

$(iv) \Rightarrow (iii)$:

In a similar way it can be proven that (4.14) implies (4.13), hence, (4.13) and (4.14) are equivalent.

$(iii) \Rightarrow (i)$:

Notice that the validity of inequalities (4.13) and (4.14) is equivalent to

$$[a_{ik}^L(\alpha); a_{ik}^R(\alpha)] \cap [a_{ij}^L(\alpha) \odot a_{jk}^L(\alpha); a_{ij}^R(\alpha) \odot a_{jk}^R(\alpha)] \ne \emptyset. \qquad (4.17)$$

Therefore, we may assume that for $i, j, k \in \{1, \cdots, n\}$ there is a a_{ik}' such that

$$a_{ik}' \in [a_{ik}^L(\alpha); a_{ik}^R(\alpha)] \cap [a_{ij}^L(\alpha) \odot a_{jk}^L(\alpha); a_{ij}^R(\alpha) \odot a_{jk}^R(\alpha)].$$

By continuity of operation $\odot$ and Weierestrass's well-known theorem, there exist $a_{ij}' \in [a_{ij}^L(\alpha); a_{ij}^R(\alpha)]$ and $a_{jk}' \in [a_{jk}^L(\alpha); a_{jk}^R(\alpha)]$ such that $a_{ik}' = a_{ij}' \odot a_{jk}'$, hence, $A' = \{a_{ij}'\}$ is consistent and, by Definition 4.2, $\tilde{A} = \{\tilde{a}_{ij}\}$ is weak α-$\odot$-consistent.

Thus, the proof of all equivalences is complete. $\qquad \square$

Remark 4.6 Properties (iii), (iv) in Proposition 4.1 are useful for checking weak α-$\odot$-consistency of FPC matrices. For a given α-$\odot$-reciprocal FPC matrix it can be easily checked whether inequality (4.13) or (4.14) is satisfied or not.

Example 4.5 Consider the additive alo-group $\mathcal{R} = (\mathbf{R}, \odot, \le)$ with $\odot = +$, see Example 4.1. Let $\tilde{A} = \{\tilde{a}_{ij}\}$ be given by triangular fuzzy number elements as follows:

$$\tilde{A} = \begin{pmatrix} (0,0,0) & (1,3,4) & (4,6,8) \\ (-4,-3,-1) & (0,0,0) & (2,4,5) \\ (-8,-6,-4) & (-5,-4,-2) & (0,0,0) \end{pmatrix},$$

or, equivalently, by α-cut notation, see the definition of fuzzy sets in Sect. 4.2. Each fuzzy set is given by the corresponding family of α-cuts, i.e., intervals. In particular, for $\alpha \in [0; 1]$, we obtain

$$\tilde{A} = \begin{pmatrix} [0; 0] & [1 + 2\alpha; 4 - \alpha] & [4 + 2\alpha; 8 - 2\alpha] \\ [-4 + \alpha; -1 - 2\alpha] & [0; 0] & [2 + 2\alpha; 5 - \alpha] \\ [-8 + 2\alpha; -4 - 2\alpha] & [-5 + \alpha; -2 - 2\alpha] & [0; 0] \end{pmatrix}.$$

Here, $\tilde{A}$ is a 3×3 matrix with triangular fuzzy number elements and the corresponding piecewise linear membership functions. Conditions (4.8) and (4.9) can be easily verified for all $\alpha \in [0; 1]$, hence, $\tilde{A}$ is $\odot$-reciprocal.

By Proposition 4.1 we check only one of the inequalities (4.13) (or (4.14)), for all triples of indices $i, j, k \in \{1, 2, 3\}$. By a simple calculation, we obtain that $\tilde{A}$ is weak α-$\odot$-consistent for all $0 \le \alpha \le \frac{5}{6}$.

It is also evident that $\tilde{A}$ is an α-$\odot$-inconsistent FPCM for $\frac{5}{6} < \alpha \le 1$.

Now, we are going to define a stronger concept of α-$\odot$-consistency (without thee adjective "weak") by formula similar to (4.5) in the crisp case. In [10], the same concept of consistency is applied to a PCM with interval elements. In the literature there exist some other concepts of consistency, e.g., Liu's consistency, approximate consistency, or strong consistency, see e.g., [24, 39], or [32]. However, these approaches are not investigated here.

Definition 4.3 Let $\alpha \in [0; 1]$. A FPC matrix $\tilde{A} = \{\tilde{a}_{ij}\}$ is said to be α-$\odot$-*consistent*, if the following condition holds:

For all $i, j, k \in \{1, ..., n\}$, it holds that

$$a_{ij}^L(\alpha) \odot a_{jk}^L(\alpha) \odot a_{ki}^L(\alpha) = a_{ik}^L(\alpha) \odot a_{kj}^L(\alpha) \odot a_{ji}^L(\alpha), \tag{4.18}$$

$$a_{ij}^R(\alpha) \odot a_{jk}^R(\alpha) \odot a_{ki}^R(\alpha) = a_{ik}^R(\alpha) \odot a_{kj}^R(\alpha) \odot a_{ji}^R(\alpha). \tag{4.19}$$

Moreover, if $\tilde{A} = \{\tilde{a}_{ij}\}$ is α-$\odot$-consistent for all $\alpha \in [0; 1]$, then $\tilde{A}$ is said to be $\odot$-*consistent*.

Example 4.6 Consider the additive alo-group $\mathscr{R} = (\mathbf{R}, \odot, \le)$ with $\odot = +$, see Example 4.1. Let $\tilde{B} = \{\tilde{b}_{ij}\}$ be given as follows:

$$\tilde{B} = \begin{pmatrix} (0, 0, 0) & (4, 6, 8) & (10, 14, 18) \\ (-7, -6, -4) & (0, 0, 0) & (6, 8, 10) \\ (-18, -14, -10) & (-11, -8, -6) & (0, 0, 0) \end{pmatrix},$$

or, equivalently, by the definition of fuzzy sets in Sect. 4.2, each fuzzy set is given by the corresponding family of α-cuts. In particular, for $\alpha \in [0; 1]$, we obtain

$$\tilde{B} = \begin{pmatrix} [0;0] & [4+2\alpha; 8-2\alpha] & [10+4\alpha; 18-4\alpha] \\ [-7+\alpha; -4-2\alpha] & [0;0] & [6+2\alpha; 10-2\alpha] \\ [-18+4\alpha; -5-2\alpha] & [-11+3\alpha; -6-2\alpha] & [0;0] \end{pmatrix}.$$

Here, $\tilde{B}$ is a 3×3 FPC matrix, particularly a FPC matrix with triangular fuzzy number elements and the corresponding piecewise linear membership functions.

Checking inequalities (4.18) and (4.19), we obtain that $\tilde{B}$ is α-$\odot$-consistent for all $0 \le \alpha \le 1$, hence, it is $\odot$-consistent.

By checking inequalities (4.13) and (4.14), it is also clear that $\tilde{B}$ is weak $\odot$-consistent FPCM. The reciprocity condition (4.9) of $\tilde{B}$ is not satisfied as, e.g.,

$$b_{21}^L(\alpha) \odot b_{12}^R(\alpha) = (-7+\alpha) + (8-2\alpha) \neq 0,$$

for all $\alpha \in [0; 1[$ which can be easily verified.

Remark 4.7 An α-$\odot$-consistent FPC matrix is not necessarily the α-$\odot$-reciprocal, as it was demonstrated in Example 4.6. In Definition 4.4, α-$\odot$-reciprocity of the FPC matrix is not assumed. In real DM problems, α-$\odot$-reciprocity condition is, however, a natural assumption. With regard to Remark 4.1, in what follows, unless otherwise stated, we shall tacitly assume that all FCPMs are $\odot$-reciprocal.

The following proposition gives a characterization of α-$\odot$-reciprocal α-$\odot$-consistent FPC matrices (see also [10] or [24, 39]).

Proposition 4.2 *Let $\alpha \in [0; 1]$, let $\tilde{A} = \{\tilde{a}_{ij}\}$ be a reciprocal FPC matrix, $[\tilde{a}_{ij}]_\alpha = [a_{ij}^L(\alpha); a_{ij}^R(\alpha)]$ be an α-cut. The following conditions are equivalent.*

(i) $\tilde{A} = \{\tilde{a}_{ij}\}$ is α-$\odot$-consistent.
(ii)   For all $i, j, k \in \{1, \cdots, n\}$

$$a_{ij}^L(\alpha) \odot a_{jk}^L(\alpha) \odot a_{ki}^L(\alpha) = a_{ik}^L(\alpha) \odot a_{kj}^L(\alpha) \odot a_{ji}^L(\alpha), \qquad (4.20)$$

is equivalent to

$$a_{ij}^R(\alpha) \odot a_{jk}^R(\alpha) \odot a_{ki}^R(\alpha) = a_{ik}^R(\alpha) \odot a_{kj}^R(\alpha) \odot a_{ji}^R(\alpha). \qquad (4.21)$$

(iii)   For all $i, j, k \in \{1, \cdots, n\}$, it holds that

$$a_{ik}^L(\alpha) \odot a_{ik}^R(\alpha) = a_{ij}^L(\alpha) \odot a_{ij}^R(\alpha) \odot a_{jk}^L(\alpha) \odot a_{jk}^R(\alpha). \qquad (4.22)$$

Proof $(i) \Rightarrow (ii)$:
Let $i, j, k \in \{1, \cdots, n\}$. By "multiplying" both sides of (4.20) by

$$a_{jk}^R(\alpha) \odot a_{ij}^R(\alpha) \odot a_{ik}^R(\alpha),$$

applying reciprocity conditions, i.e., $a_{rs}^L(\alpha) \odot a_{sr}^R(\alpha) = e$, for all $r, s \in \{1, \cdots, n\}$ and commutativity of $\odot$, we obtain (4.21).

Vice versa, by "multiplying" both sides of (4.21) by

$$a_{jk}^L(\alpha) \odot a_{ij}^L(\alpha) \odot a_{ik}^L(\alpha),$$

applying reciprocity conditions, and commutativity, we obtain (4.20). Hence (4.20) is equivalent to (4.21).

$(ii) \Rightarrow (i)$:
If (4.20) is equivalent to (4.21), then both equalities hold and $\tilde{A} = \{\tilde{a}_{ij}\}$ is α-$\odot$-consistent.

$(ii) \Rightarrow (iii)$:
Let $i, j, k \in \{1, \cdots, n\}$. "multiplying" both sides of (4.20) by

$$a_{jk}^R(\alpha) \odot a_{ij}^R(\alpha) \odot a_{ik}^R(\alpha),$$

applying reciprocity conditions, and commutativity, we obtain (4.22). Hence (4.20) implies (4.22).

$(iii) \Rightarrow (ii)$:
Let $i, j, k \in \{1, ..., n\}$. By "multiplying" both sides of (4.22) by

$$a_{ji}^L(\alpha) \odot a_{kj}^L(\alpha) \odot a_{ki}^L(\alpha)$$

and then both sides of (4.22) by

$$a_{ji}^R(\alpha) \odot a_{kj}^R(\alpha) \odot a_{ki}^R(\alpha),$$

applying reciprocity conditions, and commutativity, we obtain both (4.18) and (4.19).
Now it is proven that all three conditions (i)–(iii) are equivalent.   □

Definition 4.4 Let $\alpha \in [0; 1]$, let $\tilde{A} = \{\tilde{a}_{ij}\}$ be a FPC matrix, $[\tilde{a}_{ij}]_\alpha = [a_{ij}^L(\alpha);$ $a_{ij}^R(\alpha)]$ be an α-cut. For all $i, j \in \{1, \cdots, n\}$ denote

$$a_{ij}^m(\alpha) = (a_{ij}^L(\alpha) \odot a_{ij}^R(\alpha))^{(\frac{1}{2})}. \tag{4.23}$$

A crisp $n \times n$-matrix $A^m(\alpha) = \{a_{ij}^m(\alpha)\}$ is called an α-$\odot$-*mean matrix* associated with FPC matrix $\tilde{A} = \{\tilde{a}_{ij}\}$.

By Proposition 4.2, (iii), an FPC matrix $\tilde{A} = \{\tilde{a}_{ij}\}$ is α-$\odot$-consistent, if and only if crisp α-$\odot$-mean matrix $A^m(\alpha)$ is $\odot$-consistent, i.e., the following formula holds for all $i, j, k \in \{1, \cdots, n\}$:

$$a_{ij}^m(\alpha) = a_{ij}^m(\alpha) \odot a_{jk}^m(\alpha). \tag{4.24}$$

Moreover, if the FPC matrix $\tilde{A} = \{\tilde{a}_{ij}\}$ is α-$\odot$-consistent for all $\alpha \in [0; 1]$, then $\tilde{A}$ is $\odot$-consistent.

Remark 4.8 Notice that the α-$\odot$-consistency of FPC matrix $\tilde{A} = \{\tilde{a}_{ij}\}$ is based on the $\odot$-consistency of the associated crisp matrix, particularly, the α-$\odot$-mean matrix associated with $\tilde{A}$, with elements being generalized means (4.23) of the end points of the corresponding the α-cuts. This fact will be advantageous in deriving a corresponding priority vector of the FPC matrix as we can see in the next section.

Example 4.7 Consider the additive alo-group $\mathscr{R} = (\mathbf{R}, \odot, \leq)$ with $\odot = +$, see Example 4.1. Let $\tilde{B} = \{\tilde{b}_{ij}\}$ be given as follows:

$$\tilde{B} = \begin{pmatrix} (0, 0, 0) & (4, 6, 8) & (10, 14, 18) \\ (-8, -6, -4) & (0, 0, 0) & (6, 8, 10) \\ (-18, -14, -10) & (-10, -8, -6) & (0, 0, 0) \end{pmatrix},$$

or, equivalently, by the definition of fuzzy sets in Sect. 4.2, where each fuzzy set is given by the corresponding family of α-cuts. Particularly, for $\alpha \in [0; 1]$, we obtain

$$\tilde{B} = \begin{pmatrix} [0; 0] & [4 + 2\alpha; 8 - 2\alpha] & [10 + 4\alpha; 18 - 4\alpha] \\ [-8 + 2\alpha; -4 - 2\alpha] & [0; 0] & [6 + 2\alpha; 10 - 2\alpha] \\ [-18 + 4\alpha; -10 - 4\alpha] & [-10 + 2\alpha; -6 - 2\alpha] & [0; 0] \end{pmatrix}.$$

Evidently, $\tilde{B}$ is a FPC matrix with the corresponding piecewise linear membership functions.

Moreover, α-+-mean matrix $B^m(\alpha)$ associated with $\tilde{B}$ for $\alpha \in [0; 1]$ is calculated by (4.23) as

$$B^m(\alpha) = \begin{pmatrix} 0 & 6 & 14 \\ -6 & 0 & 8 \\ -14 & -8 & 0 \end{pmatrix}.$$

Checking equality (4.24), we obtain that $\tilde{B}$ is α-+-consistent for all $0 \leq \alpha \leq 1$, hence, it is +-consistent. Notice that $B^m(\alpha)$ is independent of α.

The following propositions give some characterizations of α-$\odot$-consistent FPC matrices, see also [10] or [24, 39]. It is a "fuzzy version" of Proposition 3.3 in the crisp case.

Proposition 4.3 *Let $\alpha \in [0; 1]$, let $\tilde{A} = \{\tilde{a}_{ij}\}$ be a FPC matrix, $[\tilde{a}_{ij}]_\alpha = [a_{ij}^L(\alpha), a_{ij}^R(\alpha)]$ be an α-cut. The following conditions are equivalent.*

(i) $\tilde{A} = \{\tilde{a}_{ij}\}$ *is α-$\odot$-consistent.*
(ii)   *There exists a vector $w(\alpha) = (w_1(\alpha), \cdots, w_n(\alpha))$ with $w_j(\alpha) \in G$, $j \in \{1, \cdots, n\}$, such that for each $i, k \in \{1, \cdots, n\}$, it holds that*

$$a_{ik}^m(\alpha) = w_i(\alpha) \div w_k(\alpha). \tag{4.25}$$

Proof Let $\alpha \in [0; 1]$.
$(i) \Rightarrow (ii)$:

Let (4.24) holds. Then for j_0, setting $w_i(\alpha) = a_{ij_0}(\alpha) \div a_{j_0k}(\alpha)$, by (4.23) and reciprocity we obtain (4.25).

$(ii) \Rightarrow (i)$:

Let (4.25) holds, then

$$(a_{ik}^L(\alpha) \odot a_{ik}^R(\alpha))^{(\frac{1}{2})} = (w_i(\alpha) \div w_j(\alpha) \odot w_j(\alpha) \div w_k(\alpha))^{(\frac{1}{2})} =$$

$$= (a_{ij}^L(\alpha) \odot (a_{ij}^R(\alpha))^{(\frac{1}{2})} \odot (a_{jk}^L(\alpha) \odot a_{jk}^R(\alpha))^{(\frac{1}{2})},$$

hence, we obtain (4.24). □

Proposition 4.4 *Let $\alpha \in [0; 1]$, let $\tilde{A} = \{\tilde{a}_{ij}\}$ be a FPC matrix.*

If $\tilde{A}$ is α-$\odot$-consistent then $\tilde{A}$ is weak α-$\odot$-consistent.

Moreover, if $\tilde{A}$ is $\odot$-consistent then $\tilde{A}$ is weak $\odot$-consistent.

Proof Let $i, j, k \in \{1, ..., n\}$. If $\tilde{A}$ is α-$\odot$-consistent then by (4.23) and (4.24) we obtain

$$(a_{ij}^L(\alpha) \odot a_{jk}^L(\alpha)) \odot (a_{ij}^R(\alpha) \odot a_{jk}^R(\alpha)) = a_{ik}^L(\alpha) \odot a_{ik}^R(\alpha). \qquad (4.26)$$

Assuming $a_{ik}^R(\alpha) < a_{ij}^L(\alpha) \odot a_{jk}^L(\alpha)$, then by (4.26) we obtain

$$a_{ik}^R(\alpha) \odot (a_{ij}^R(\alpha) \odot a_{jk}^R(\alpha)) < a_{ik}^L(\alpha) \odot a_{ik}^R(\alpha),$$

hence, $a_{ij}^R(\alpha) \odot a_{jk}^R(\alpha) < a_{ik}^L(\alpha)$, therefore,

$$a_{ik}^R(\alpha) < a_{ij}^L(\alpha) \odot a_{jk}^L(\alpha) \le a_{ij}^R(\alpha) \odot a_{jk}^R(\alpha) < a_{ik}^L(\alpha),$$

a contradiction to $a_{ik}^L(\alpha) \le a_{ik}^R(\alpha)$.

Hence, it holds $a_{ik}^R(\alpha) \ge a_{ij}^L(\alpha) \odot a_{jk}^L(\alpha)$, which is (4.14). Hence, by Proposition 4.1, (iv), $\tilde{A}$ is weak α-$\odot$-consistent.

The last part of the proposition is evident. □

Example 4.8 Let $\tilde{A} = \{\tilde{a}_{ij}\}$ be the FPCM from Example 4.5 given by the corresponding family of α-cuts, i.e., intervals. In particular, for $\alpha \in [0; 1]$, we obtain

$$\tilde{A} = \begin{pmatrix} [0; 0] & [1 + 2\alpha; 4 - \alpha] & [4 + 2\alpha; 8 - 2\alpha] \\ [-4 + \alpha; -1 - 2\alpha] & [0; 0] & [2 + 2\alpha; 5 - \alpha] \\ [-8 + 2\alpha; -4 - 2\alpha] & [-5 + \alpha; -2 - 2\alpha] & [0; 0] \end{pmatrix}.$$

Here, $\tilde{A}$ is a FPC matrix with triangular fuzzy number elements and the corresponding piecewise linear membership functions.

In Example 4.5 we obtained that $\tilde{A}$ is weak α-+-consistent for all $0 \le \alpha \le \frac{5}{6}$.

Moreover, it can be easily verified that $\tilde{A}$ is not an α-+-consistent FPCM for any α, $0 < \alpha \le 1$, as condition (4.24) is not satisfied. In particular, for $0 < \alpha \le 1$ we obtain

$$(a_{13}^L(\alpha) + a_{13}^R(\alpha))/2 \neq (a_{12}^L(\alpha) + a_{12}^R(\alpha))/2 + (a_{23}^L(\alpha) + a_{23}^R(\alpha))/2,$$

as

$$(4 + 2\alpha + 8 - 2\alpha)/2 = 6 \neq (1 + 2\alpha + 4 - \alpha)/2 + (2 + 2\alpha + 5 - \alpha)/2 = 6 + \alpha.$$

4.4 Desirable Properties of the Priority Vector

Pairwise comparisons matrices may violate some desirable conditions of multiple criteria decision-making: the "best" alternative with respect to DM's preferences is selected from the set of non-dominated alternatives, on condition that this set is nonempty. The other PCMs may violate the condition of order of preferences (the so called COP condition), or the preservation of the (intensity of) preference (the so called POP/POIP condition), see Bana e Costa and Vansnick [2].

As it was already mentioned in Chap. 3, Sect. 3.2, a PC matrix with crisp elements $A = \{a_{ij}\}$ is said to satisfy the preservation of order preference condition (POP condition) with respect to priority vector w if

$$a_{ij} > e \Rightarrow w_i > w_j. \tag{4.27}$$

A PC matrix A is said to satisfy the preservation of order intensity preference condition (POIP condition) with respect to vector w if

$$a_{ij} > e, a_{kl} > e, \text{ and } a_{ij} > a_{kl} \Rightarrow w_i \div w_j > w_k \div w_l. \tag{4.28}$$

And, finally, a PC matrix A is said to satisfy the reliable preference (RP) condition with respect to priority vector w if

$$a_{ij} > e \Rightarrow w_i > w_j, \tag{4.29}$$

$$a_{ij} = e \Rightarrow w_i = w_j. \tag{4.30}$$

From (4.29) in the above definition, it is evident that if a crisp PC matrix A satisfies the RP condition with respect to priority vector w, then A satisfies the POP condition with respect to priority vector w. The opposite is not true.

Because of the fuzzy character of elements in FPC matrices, we shall not investigate the POIP condition in more detail in what follows.

Let $A = \{a_{ij}\}$ be a crisp consistent PC matrix, and let $w = (w_1, \cdots, w_n)$ be a priority vector associated with A satisfying (4.6). Then it is obvious that the FS, POP, POIP, and RP conditions are satisfied. Moreover, it is well known (see e.g., [35]), that for each crisp consistent PC matrix, the priority vector satisfying (4.6) can be generated either by the eigenvalue method (EVM) or by the geometric mean method (GMM).

Now, we are going to define the POP condition for a FPC matrix $\tilde{A} = \{\tilde{a}_{ij}\}$, as we mentioned before, the POIP condition will not be dealt with, here. We start with the definition of domination between two alternatives.

Definition 4.5 Let $c_i, c_j \in \mathscr{C}$, $\tilde{A} = \{\tilde{a}_{ij}\}$ be an FPC matrix on the alo-group $\mathscr{G} = (G, \odot, \leq), \alpha \in [0; 1]$.

(i) c_i *α-weakly dominates* c_j, if there exists $a_{ij} \in [\tilde{a}_{ij}]_\alpha$, such that it holds that $a_{ij} > e$.
       Moreover, c_i *weakly dominates* c_j, if c_i α-weakly dominates c_j, for all $\alpha \in [0; 1]$.

(ii) c_i *α-strongly dominates* c_j, if for each $a_{ij} \in [\tilde{a}_{ij}]_\alpha$, it holds that $a_{ij} > e$.
       Moreover, c_i *strongly dominates* c_j, if c_i α-strongly dominates c_j, for all $\alpha \in [0; 1]$.

(iii) c_i *α-mean dominates* c_j, if $a_{ij}^m(\alpha) > e$, where

$$a_{ij}^m(\alpha) = (a_{ij}^L(\alpha) \odot a_{ij}^R(\alpha))^{(\frac{1}{2})}. \tag{4.31}$$

Moreover, c_i *mean dominates* c_j, if c_i α-mean dominates c_j, for all $\alpha \in [0; 1]$.

The following properties are evident, the proofs are left to the reader.

Proposition 4.5 *Let* $c_i, c_j \in \mathscr{C}$, $\tilde{A} = \{\tilde{a}_{ij}\}$ *be an FPC matrix on alo-group* $\mathscr{G}, \alpha \in [0; 1]$, *let* $[\tilde{a}_{ij}]_\alpha = [a_{ij}^L(\alpha); a_{ij}^R(\alpha)]$ *be an* α-*cut.*

(i) c_i α-*weakly dominates* c_j *if and only if* $a_{ij}^R(\alpha) > e$.
(ii) c_i *weakly dominates* c_j *if and only if* $a_{ij}^R(1) > e$.
(iii) c_i α-*strongly dominates* c_j *if and only if* $a_{ij}^L(\alpha) > e$.
(iv) c_i *strongly dominates* c_j *if and only if* $a_{ij}^L(0) > e$.
(v) c_i α-*mean dominates* c_j *if and only if* $a_{ij}^m(\alpha) > e$.
(vi) c_i *mean dominates* c_j *if and only if for all* $\alpha \in [0; 1] : a_{ij}^m(\alpha) > e$.

Now, we define the weak POP and strong POP conditions depending on the weak and strong domination of alternatives.

Definition 4.6 Let $c_i, c_j \in \mathscr{C}$, $\tilde{A} = \{\tilde{a}_{ij}\}$ be a FPC matrix on the alo-group $\mathscr{G} = (G, \odot, \leq)$, $w = (w_1, w_2, \cdots, w_n)$, $w_i \in G$, be a priority vector, $\alpha \in [0; 1]$.

(i)   We say that the *α-weak preservation of order preference condition (α-WPOP condition)* is satisfied:
       if c_i α-weakly dominates c_j, then $w_i > w_j$.

(ii)   We say that the *weak preservation of order preference condition (WPOP condition)* is satisfied:
       if c_i weakly dominates c_j, then $w_i > w_j$.

(iii) We say that the *α-strong preservation of order preference condition (α-SPOP condition)* is satisfied:
       if c_i α-strongly dominates c_j, then $w_i > w_j$.

(iv) We say that the *strong preservation of order preference condition (SPOP condition)* is satisfied:

if c_i strongly dominates c_j, then $w_i > w_j$.

(v) We say that the *α-mean preservation of order preference condition (α-MPOP condition)* is satisfied:

if c_i α-mean dominates c_j, then $w_i > w_j$.

(vi) We say that the *mean preservation of order preference condition (MPOP condition)* is satisfied:

if c_i mean dominates c_j, then $w_i > w_j$.

Remark 4.9 Notice, that the concepts of the satisfaction of the α-weak, α-strong, and/or α-mean preservation of order preference condition (i.e., α-WPOP, α-SPOP, resp. α-MPOP condition) is introduced for a given FPC matrix $\tilde{A}$ and given priority vector w. If $\tilde{A} = \{\tilde{a}_{ij}\}$ is a crisp FPC matrix, then the α-WPOP, WPOP, α-SPOP, SPOP, α-MPOP, and MPOP conditions coincide with the usual POP condition.

Remark 4.10 By Definition 2.6, the WPOP condition implies the α-WPOP condition for any $\alpha \in [0; 1]$. Analogically, SPOP and MPOP condition implies α-SPOP and α-MPOP condition for any $\alpha \in [0; 1]$.

Remark 4.11 It is evident, that the α-SPOP condition implies α-WPOP condition for all $\alpha \in [0; 1]$, but not vice versa. Similarly, the α-SPOP condition implies α-MPOP condition and α-MPOP conditions imply the α-WPOP condition for all $\alpha \in [0; 1]$, but not vice versa.

Here, we assume that all FPC matrices are α-$\odot$ reciprocal see Definition 4.3. In practice, this assumption is not restrictive, see Remark 4.1.

In what follows we introduce the local error indexes and global error index based on the $\odot$-mean of the end points of α-cuts of the fuzzy elements of the FPCM. Here, we apply a similar approach as in Chap. 3, Sect. 3.2.

Definition 4.7 Let $\tilde{A} = \{\tilde{a}_{ij}\}$ be an FPC matrix on alo-group $\mathcal{G} = (G, \odot, \leq)$. For each pair $i, j \in \{1, \ldots, n\}$, and a priority vector $w = (w_1, w_2, \cdots, w_n)$, $w_i \in G$, $\alpha \in [0; 1]$ and for $a_{ij}^m(\alpha) = (a_{ij}^L(\alpha) \odot a_{ij}^R(\alpha))^{(\frac{1}{2})}$ let us denote

$$\varepsilon^m(i, j, w, \alpha) = a_{ij}^m(\alpha) \odot w_j \div w_i, \tag{4.32}$$

Further, let

$$\varepsilon^m(i, j, w, \alpha) = \max\{\varepsilon^m(i, j, w, \alpha), (\varepsilon^m(i, j, w, \alpha))^{(-1)}\}, \tag{4.33}$$

Moreover, define the $n \times n$ matrix of *local error indexes* $\varepsilon^m(w, \alpha)$ as

$$\varepsilon^m(w, \alpha) = \{\varepsilon^m(i, j, w, \alpha)\}. \tag{4.34}$$

Definition 4.8 Let $\alpha \in [0; 1]$. The *global error index* $\mathscr{E}(\tilde{A}, w, \alpha)$, for a FPC matrix $\tilde{A} = \{\tilde{a}_{ij}\}$ and a priority vector $w = (w_1, \ldots, w_n)$ is defined as the maximal element of matrix of local errors $\varepsilon^m(w, \alpha)$, i.e.,

$$\mathscr{E}(\tilde{A}, w, \alpha) = \max_{i,j \in \{1,\ldots,n\}} \varepsilon^m(i, j, w, \alpha). \tag{4.35}$$

Remark 4.12 Let $\tilde{A} = \{\tilde{a}_{ij}\}$ be a FPC matrix and $w = (w_1, \ldots, w_n)$ be a priority vector, $\alpha \in [0; 1]$. It is clear from Definition 4.8 and (4.33) that

$$\mathscr{E}(\tilde{A}, w, \alpha) \geq e. \tag{4.36}$$

Lemma 4.1 *Let* $\tilde{A} = \{\tilde{a}_{ij}\}$ *be an FPC matrix. Then for each* $i, j \in \{1, \ldots, n\}$, *and priority vector* $w = (w_1, \cdots, w_n)$, $w_i \in G$, $\alpha \in [0; 1]$:

$$\varepsilon^m(i, j, w, \alpha) \odot w_i \div w_j \geq a_{ij}^m(\alpha), \tag{4.37}$$

Proof Let $i, j \in \{1, \ldots, n\}$. By Definition 4.8, particularly (4.32) and (4.33), we obtain

$$\varepsilon^m(i, j, w, \alpha) \geq a_{ij}^m(\alpha) \odot w_j \div w_i.$$

Multiplying (using operation $\odot$) both sides by $w_i \div w_j$ we get

$$\varepsilon^m(i, j, w, \alpha) \odot w_i \div w_j \geq a_{ij}^m(\alpha).$$

Hence, we have formula (4.37). $\qquad\qquad\qquad\qquad\qquad\qquad\qquad\qquad\square$

Proposition 4.6 *Let* $\tilde{A} = \{\tilde{a}_{ij}\}$ *be an FPC matrix,* $\alpha \in [0; 1]$.

(i) *If* $\tilde{A}$ *is weak* α-$\odot$-*consistent, then there exists a priority vector* $w(\alpha) = (w_1(\alpha), w_2(\alpha), \ldots, w_n(\alpha))$, *such that the* α-*SPOP condition is satisfied, i.e.,*

$$a_{ij}^L(\alpha) > e \text{ implies } w_i(\alpha) > w_j(\alpha).$$

If $\tilde{A}$ *is weak* $\odot$-*consistent, then there exists a priority vector* $w = (w_1, w_2, \ldots, w_n)$, *such that the SPOP condition is satisfied, i.e.,*

$$a_{ij}^L(0) > e \text{ implies } w_i > w_j.$$

(ii) *If* $\tilde{A}$ *is* α-$\odot$-*consistent, then there exists a priority vector* $w(\alpha) = (w_1(\alpha), \cdots, w_n(\alpha))$, *such that the* α-*MPOP condition is satisfied, i.e.,*

$$a_{ij}^m(\alpha) > e \text{ implies } w_i(\alpha) > w_j(\alpha).$$

Moreover, if $\tilde{A}$ is $\odot$-consistent, then there exists a priority vector $w = (w_1, \cdots, w_n)$, such that the MPOP condition is satisfied, i.e.,

$$a_{ij}^m(1) > e \text{ implies } w_i > w_j.$$

Proof (i) Suppose that $\tilde{A}$ is weak α-$\odot$-consistent, let

$$a_{ij}^L(\alpha) > e. \tag{4.38}$$

By Proposition 4.1 there exists a priority vector $w(\alpha) = (w_1(\alpha), \cdots, w_n)(\alpha)$, such that by (4.12)

$$a_{ij}^L(\alpha) \leq w_i(\alpha) \div w_j(\alpha), \tag{4.39}$$

hence, by (4.38) and (4.39), $w_i(\alpha) \div w_j(\alpha) > e$, which is equivalent to $w_i(\alpha) > w_j(\alpha)$ and the α-SPOP condition is satisfied. The rest of the proposition follows from Definition 4.6.

(ii) Suppose that $\tilde{A}$ is α-$\odot$-consistent, let

$$a_{ij}^m(\alpha) > e. \tag{4.40}$$

By Proposition 4.1 there exists a priority vector $w(\alpha) = (w_1(\alpha), \ldots, w_n(\alpha))$, such that

$$a_{ij}^m(\alpha) = w_i(\alpha) \div w_j(\alpha), \tag{4.41}$$

hence, by (4.40) and (4.41), $w_i(\alpha) \div w_j(\alpha) > e$, which is equivalent to $w_i(\alpha) > w_j(\alpha)$ and α-MPOP condition is satisfied. The rest of the proposition follows from Definition 4.6. $\qquad\qquad\square$

Proposition 4.7 *Let $\tilde{A} = \{\tilde{a}_{ij}\}$ be a FPC matrix, $w = (w_1, w_2, \cdots, w_n)$ be a priority vector, $\alpha \in [0; 1]$, $i, j \in \{1, \cdots, n\}$.*

(i)   *If $a_{ij}^L(\alpha) > \varepsilon^m(i, j, w, \alpha)$, then α-SPOP condition is satisfied, i.e.,*
      $a_{ij}^L(\alpha) > e$ implies $w_i > w_j$.
(ii) *If $a_{ij}^m(\alpha) > \varepsilon^m(i, j, w, \alpha)$, then α-MPOP condition is satisfied, i.e.,*
      $a_{ij}^m(\alpha) > e$ implies $w_i > w_j$.

Proof (i) Suppose that

$$a_{ij}^L(\alpha) > \varepsilon^m(i, j, w, \alpha). \tag{4.42}$$

Applying operation $\odot$ to both sides of inequality (4.42) by $w_i \div w_j$, and using Lemma 4.1, we obtain

$$a_{ij}^L(\alpha) \odot w_i \div w_j > \varepsilon^m(i, j, w, \alpha) \odot w_i \div w_j \geq a_{ij}^m(\alpha) \geq a_{ij}^L(\alpha).$$

From the left- and the right-hand side of this formula, we get $w_i \div w_j > e$, which is equivalent to $w_i > w_j$. Hence, the α-SPOP condition is satisfied.

(ii) The α-MPOP condition can be proven analogically:
Suppose that

$$a_{ij}^m(\alpha) > \varepsilon^m(i, j, w, \alpha). \tag{4.43}$$

Applying operation $\odot$ to both sides of inequality (4.43) by $w_i \div w_j$, and using Lemma 4.1, we obtain

$$a_{ij}^m(\alpha) \odot w_i \div w_j > \varepsilon^m(i, j, w, \alpha) \odot w_i \div w_j \geq a_{ij}^m(\alpha).$$

From the left- and the right-hand side of this formula, we get $w_i \div w_j > e$, which is equivalent to $w_i > w_j$. Hence, the α-MPOP condition is satisfied. $\square$

A sufficient condition for satisfying the POP condition formulated in Proposition 4.7 is based on the local error index from Definition 4.7. A similar sufficient condition can be also formulated for the global error index (4.35), from Definition 4.8.

Proposition 4.8 *Let $\tilde{A} = \{\tilde{a}_{ij}\}$ be an FPC matrix, $w = (w_1, w_2, \cdots, w_n)$ be a priority vector, $i, j \in \{1, \cdots, n\}$, $\alpha \in [0; 1]$.*

(i)   *If $a_{ij}^L(\alpha) > \mathscr{E}(\tilde{A}, w, \alpha)$, then the α-SPOP condition is satisfied, i.e.,*
       $a_{ij}^L(\alpha) > e$ implies $w_i > w_j$.

(ii)   *If $a_{ij}^m(\alpha) > \mathscr{E}(\tilde{A}, w, \alpha)$, then the α-MPOP condition is satisfied, i.e.,*
       $a_{ij}^m(\alpha) > e$ implies $w_i > w_j$.

Proof (i) The α-SPOP condition follows directly from Proposition 4.7 and from the following inequality, see (4.35),

$$\mathscr{E}(\tilde{A}, w, \alpha) \geq \varepsilon^m(i, j, w, \alpha). \tag{4.44}$$

(ii) The α-MPOP condition follows directly from Proposition 4.7 and from the following inequality

$$\mathscr{E}(\tilde{A}, w, \alpha) \geq \varepsilon^m(i, j, w, \alpha). \tag{4.45}$$

$\square$

4.5 Priority Vectors

In this section we extend our considerations from Chap. 3, Sect. 3.3, from crisp PC matrix $A = \{a_{ij}\}$ to FPC matrix $\tilde{A} = \{\tilde{a}_{ij}\}$ by using crisp α-$\odot$-mean matrix $A^m(\alpha) = \{a_{ij}^m(\alpha)\}$ associated with FPC matrix $\tilde{A} = \{\tilde{a}_{ij}\}$. Here, the elements of crisp PC matrix $A^m(\alpha)$ are defined by (4.23) as

$$a_{ij}^m(\alpha) = (a_{ij}^L(\alpha) \odot a_{ij}^R(\alpha))^{(\frac{1}{2})},$$

depending on the previously given $\alpha \in [0; 1]$.

Now, we propose a method for calculating the priority vector of $n \times n$ FPC matrix $\tilde{A} = \{\tilde{a}_{ij}\}$ for the purpose of rating the alternatives $c_1, \cdots, c_n \in \mathscr{C}$. As it has been stated before, we do not follow the way of calculating the *fuzzy* priority vector proposed, e.g., in [13, 15] and others. Here, we generate a crisp priority vector, therefore, so no defuzzification will be necessary for final ranking of the alternatives.

The proposed method for calculating the priority vector can be divided into two steps as follows.

Step 1.

In *Step 1* we check whether the given FPC matrix $\tilde{A} = \{\tilde{a}_{ij}\}$ is weak α-$\odot$-consistent for some α, where $0 \leq \alpha \leq 1$. Then we calculate the maximal such α denoted by α^*. By Remark 4.4, FPCM $\tilde{A}$ is therefore weak α-$\odot$-consistent for all $\alpha \leq \alpha^*$. The following optimization problem is solved:

(P1)

$$\alpha \longrightarrow \max; \tag{4.46}$$

subject to

$$a_{ij}^L(\alpha) \leq w_i \div w_j \leq a_{ij}^R(\alpha) \text{ for all } i, j \in \{1, \cdots, n\}, \tag{4.47}$$

$$\bigodot_{k=1}^{n} w_k = e, \tag{4.48}$$

$$0 \leq \alpha \leq 1, w_k \in G, \text{ for all } k \in \{1, \cdots, n\}. \tag{4.49}$$

In problem (P1), with variables $\alpha, w_1, \ldots, w_n$, the objective function (4.46) is maximized under the constraints securing that FPC matrix $\tilde{A} = \{\tilde{a}_{ij}\}$ is weak α-$\odot$-consistent, in (4.47), and $w = (w_1, \cdots, w_n)$, in (4.48), is normalized.

If optimization problem (P1) has a feasible solution, i.e., system of constraints (4.47)–(4.49) also has a solution, then (P1) has also an optimal solution. Let α^* and $w^1 = (w_1^1, \cdots, w_n^1)$ be an optimal solution of problem (P1). Then α^* is called the *weak $\odot$-consistency grade of FPC matrix $\tilde{A}$*, denoted by $g_\odot(\tilde{A})$, i.e., we define

$$g_\odot(\tilde{A}) = \alpha^*. \tag{4.50}$$

Here, $0 \leq \alpha^* \leq 1$. Moreover, if $g_\odot(\tilde{A}) = 1$, then FPC matrix $\tilde{A}$ is weak $\odot$-consistent.

If optimization problem (P1) has no feasible solution, which means that FPC matrix $\tilde{A} = \{\tilde{a}_{ij}\}$ is weak α-$\odot$-consistent for no $\alpha \in [0; 1]$, then we define

$$g_\odot(\tilde{A}) = 0. \tag{4.51}$$

In that case, the corresponding priority vector will be defined below.

Go to Step 2.

Remark 4.13 In general, problem (P1) is a nonlinear optimization problem that may be solved by some numerical method, e.g., by the well-known dichotomy method, which is a sequence of relatively simple optimization problems (see e.g., [6]).

In the next step we are going to obtain a corresponding priority vector with our desirable properties.

Step 2.
First, assume that problem (P1) is *feasible*. By solving the new optimization problem with α^*-$\odot$-mean matrix $A^* = \{a_{ij}^*\}$ associated with FPC matrix $\tilde{A} = \{\tilde{a}_{ij}\}$, see (4.23), we obtain a corresponding priority vector, eventually with our desirable properties FS, POP, and RP. Here, the elements of crisp PC matrix $A^* = \{a_{ij}^*\}$ are defined by (4.23) as

$$a_{ij}^* = (a_{ij}^L(\alpha^*) \odot a_{ij}^R(\alpha^*))^{(\frac{1}{2})},$$

where $\alpha^* \in [0; 1]$ has been calculated in Step 1, $\alpha^* = g_\odot(\tilde{A})$.

Now, we solve problem (P2) as follows. Let $A^* = \{a_{ij}^*\} \in PC_n(\mathscr{G})$ be a PC matrix, $\varepsilon > e$ be a suitable constant. Based on this PCM, we define the following two sets of indexes:

$$I^{(1)}(A^*) = \{(i, j)|i, j \in \{1, \cdots, n\}, a_{ij}^* = e\}, \tag{4.52}$$

$$I^{(2)}(A^*) = \{(i, j)|i, j \in \{1, \cdots, n\}, a_{ij}^* > e\}, \tag{4.53}$$

An *error index*, $\mathscr{E}(A^*, w)$, of $A^* = \{a_{ij}^*\}$ and $w = (w_1, \ldots, w_n)$ has already been defined by (4.35) as

$$\mathscr{E}(A^*, w, \alpha^*) = \max\{\|a_{ij}^* \odot w_j \div w_i\| | i, j \in \{1, \cdots, n\}\}. \tag{4.54}$$

(P2)
$$\mathscr{E}(A^*, w, \alpha^*) \longrightarrow \min; \tag{4.55}$$

subject to

$$a_{ij}^L(\alpha^*) \le w_i \div w_j \le a_{ij}^R(\alpha^*) \text{ for all } i, j \in \{1, \cdots, n\}, \tag{4.56}$$

$$\bigodot_{k=1}^n w_k = e, w_k \in G \quad \text{for all } k \in \{1, \cdots, n\}. \tag{4.57}$$

Problem (P2) with variables $w_1, \ldots, w_n$ is feasible, as $w^1 = (w_1^1, \ldots, w_n^1)$ is a part of the feasible solution of problem (P1), hence, it is also a feasible solution of (P2) with the objective function (4.55).

The optimal solution $w^* = (w_1^*, \cdots, w_n^*)$ of (P2) will be called the $\odot$-*priority vector of* $\tilde{A}$.

Second, assume that problem (P1) is *infeasible*. By solving problem (P2), now without constraints (4.56), and, with $\alpha^* = 0$, where α^*-$\odot$-mean matrix $A^* = \{a_{ij}^*\}$ is associated with FPC matrix $\tilde{A} = \{\tilde{a}_{ij}\}$, see (4.23), we obtain a corresponding priority vector $w^* = (w_1^*, \cdots, w_n^*)$, eventually with desirable properties FS, POP, and RP, i.e., (4.58), (4.59). Here, however, the elements of crisp PC matrix $A^* = \{a_{ij}^*\}$ are defined by (4.23) as

$$a_{ij}^* = (a_{ij}^L(0) \odot a_{ij}^R(0))^{(\frac{1}{2})}.$$

In order to secure that an $\odot$-priority vector satisfies the desirable properties FS, POP, and RP, we have to solve problem (P2) with two additional constraints:

$$w_r = w_s \ \forall (r, s) \in I^{(1)}(A^*), \tag{4.58}$$

$$w_r \geq w_s \odot \varepsilon \ \forall (r, s) \in I^{(2)}(A^*). \tag{4.59}$$

The existence of such optimal solution satisfying properties (4.58) and (4.59), i.e., a priority vectors satisfying the desirable properties is, however, not secured.

Remark 4.14 In general, the uniqueness of the optimal solution of (P2) is not preserved. Depending on the particular group operation $\odot$, problem (P2) may have multiple optimal solutions which is an unfavorable property from the point of view of the DM. In this case, the DM should reconsider some (fuzzy) evaluations in the original pairwise comparison matrix.

4.6 Measuring Inconsistency of FPC Matrices

If for some $\alpha \in [0; 1]$ the FPC matrix $\tilde{A} = \{\tilde{a}_{ij}\}$ is not weak α-$\odot$-consistent, then $\tilde{A}$ is called to be α-$\odot$-*inconsistent*. It is useful to introduce a concept of measuring inconsistency by an inconsistency index.

Let $\tilde{A} = \{\tilde{a}_{ij}\}$, be a FPCM and let $A = \{a_{ij}\}$ denote a PCM with the elements from 0-cut of $\tilde{A} = \{\tilde{a}_{ij}\}$, i.e., $a_{ij} \in [\tilde{a}_{ij}]_0, i, j \in \{1, \cdots, n\}$.

The $\odot$-*inconsistency index of* $\tilde{A}$ $I_{\odot}(\tilde{A})$ is defined as follows.

$$I_{\odot}(\tilde{A}) = \inf\{\sup\{\mathscr{E}(A, w, 0) | a_{ij} \in [\tilde{a}_{ij}]_0, i, j \in \{1, \cdots, n\}\} | \bigodot_{k=1}^{n} w_k = e\}. \tag{4.60}$$

Remark 4.15 In particular, if $\tilde{A} = \{a_{ij}\}$ is a crisp FPC matrix, then $\odot$-inconsistency index $I_{\odot}(\tilde{A}) = e$, if and only if $A = \tilde{A}$ is (weak) $\odot$-consistent.

Proposition 4.9 *If $\tilde{A} = \{\tilde{a}_{ij}\}$ is a FPC matrix, then exactly one of the following two cases occurs:*

- *Problem (P1) has a feasible solution α^*. Then weak consistency grade $g_{\odot}(\tilde{A}) = \alpha^*$, $0 \leq \alpha^* \leq 1$.*
  For each α, such that $0 \leq \alpha \leq \alpha^* \leq 1$, FPC matrix $\tilde{A}$ is weak α-$\odot$-consistent. The associated priority vector $w^* = (w_1^*, ..., w_n^*)$ is the optimal solution of (P2) and $I_{\odot}(\tilde{A}) \geq e$.

- *Problem (P1) has no feasible solution. Then consistency grade* $g_\odot(\tilde{A}) = 0$, $\tilde{A}$ *is $\odot$- inconsistent, hence* $I_\odot(\tilde{A}) > e$. *The associated priority vector* $w^* = (w_1^*, ..., w_n^*)$ *is the optimal solution of (P2) with* $A^* = \{a_{ij}^*\}$ *and*

$$a_{ij}^* = (a_{ij}^L(0) \odot a_{ij}^R(0))^{(\frac{1}{2})}.$$

The following three examples demonstrate various consistency properties of PCMs: $\tilde{A}_1$ is weak $\odot$-consistent for all $0 < \alpha < 1$, whereas $\tilde{A}_2$ is weak $\odot$-consistent for all $0 < \alpha \leq \alpha^* < 1$, and $\tilde{A}_3$ is weak $\odot$-inconsistent for all $0 < \alpha < 1$, here, $\odot = \cdot$.

Example 4.9 Consider the multiplicative alo-group $\mathcal{R}_+ = (\mathbf{R}_+, \cdot, \leq)$ with $\odot = \cdot$ (see Example 4.2). Let the three alternatives $\mathscr{C} = \{c_1, c_2, c_3\}$, FPCM $\tilde{A}_1 = \{\tilde{a}_{1ij}\}$ be given by triangular fuzzy number elements as follows:

$$\tilde{A}_1 = \begin{pmatrix} (1,1,1) & (1,2,3) & (7,8,9) \\ (\frac{1}{3},\frac{1}{2},1) & (1,1,1) & (3,4,5) \\ (\frac{1}{9},\frac{1}{8},\frac{1}{7}) & (\frac{1}{5},\frac{1}{4},\frac{1}{3}) & (1,1,1) \end{pmatrix},$$

or, equivalently, by α-cut notation. Each fuzzy set is given by the corresponding family of α-cuts, i.e., intervals. In particular, for $\alpha \in [0; 1]$, we obtain

$$\tilde{A}_1 = \begin{pmatrix} [1; 1] & [1 + \alpha; 3 - \alpha] & [7 + \alpha; 9 - \alpha] \\ [\frac{1}{3-\alpha}; \frac{1}{1+\alpha}] & [1; 1] & [3 + \alpha; 5 - \alpha] \\ [\frac{1}{9-\alpha}; \frac{1}{7+\alpha}] & [\frac{1}{5-\alpha}; \frac{1}{3+\alpha}] & [1; 1] \end{pmatrix}.$$

Here, $\tilde{A}_1$ is a 3×3 PC matrix with triangular fuzzy number elements and the corresponding piecewise linear membership functions.

The priority vector of $\tilde{A}_1$ is obtained as the optimal solution of problem (P2), particularly, $w^* = (w_1^*, w_2^*, w_3^*) = (2.520, 1.260, 0.315)$.

The corresponding ranking of alternatives is $c_1 > c_2 > c_3$.

The weak consistency grade is $g(\tilde{A}) = \alpha^* = 1.0$. Hence, $\tilde{A}_1$ is weak α-$\cdots$-consistent for all $0 \leq \alpha \leq 1$ or in other words, $\tilde{A}$ is weak $\cdots$-consistent.

Inconsistency index $I.(\tilde{A}_1) = \mathscr{E}(A_1^*, w^*, 1) = 1.000$.

Example 4.10 Consider the multiplicative alo-group $\mathcal{R}_+ = (\mathbf{R}_+, ., \leq)$ with $\odot = \cdot$, see Example 4.2. Let $\tilde{A}_2 = \{\tilde{a}_{2ij}\}$ be given by triangular fuzzy number elements as follows:

$$\tilde{A}_2 = \begin{pmatrix} (1,1,1) & (1,2,3) & (7,8,9) \\ (\frac{1}{3},\frac{1}{2},1) & (1,1,1) & (2,3,4) \\ (\frac{1}{9},\frac{1}{8},\frac{1}{7}) & (\frac{1}{4},\frac{1}{3},\frac{1}{2}) & (1,1,1) \end{pmatrix},$$

or, equivalently, by α-cut notation, for $\alpha \in [0; 1]$, we obtain

$$\tilde{A}_2 = \begin{pmatrix} [1;1] & [1+\alpha; 3-\alpha] & [7+\alpha; 9-\alpha] \\ [\frac{1}{3-\alpha}; \frac{1}{1+\alpha}] & [1;1] & [2+\alpha; 4-\alpha] \\ [\frac{1}{9-\alpha}; \frac{1}{7+\alpha}] & [\frac{1}{4-\alpha}; \frac{1}{2+\alpha}] & [1;1] \end{pmatrix}.$$

Here, $\tilde{A}_2$ is a 3×3 matrix with triangular fuzzy number elements and the corresponding piecewise linear membership functions.

By solving problem (P1), we obtain the weak consistency grade $g.(A_2) = 0.683$. Hence, $\tilde{A}_2$ is weak α-$\cdots$-consistent for all $0 \leq \alpha \leq 0.683$.

The priority vector of $\tilde{A}_2$ is obtained as the optimal solution of problem (P2), particularly, $w^{**} = (w_1^{**}, w_2^{**}, w_3^{**}) = (2.509, 1.147, 0.347)$.

The corresponding ranking of alternatives is $c_1 > c_2 > c_3$.

Inconsistency index $I.(\tilde{A}_2) = \mathscr{E}(A_2^*, w^{**}, 0.683) = 1.126 > 1$, hence, $\tilde{A}_2$ is $\cdots$ inconsistent.

It is evident that the 0.683-MPOP condition as well as the 0.683-MRP condition is met.

Example 4.11 Consider the multiplicative alo-group $\mathscr{R}_+ = (\mathbf{R}_+, \cdot, \leq)$ with $\odot = \cdot$ (see Example 4.2). Let $\tilde{A}_3 = \{\tilde{a}_{3ij}\}$ be given by triangular fuzzy number elements as

$$\tilde{A}_3 = \begin{pmatrix} (1,1,1) & (1, \frac{3}{2}, 3) & (7,8,9) \\ (\frac{1}{3}, \frac{2}{3}, 1) & (1,1,1) & (2, \frac{5}{2}, 3) \\ (\frac{1}{9}, \frac{1}{8}, \frac{1}{7}) & (\frac{1}{3}, \frac{2}{5}, \frac{1}{2}) & (1,1,1) \end{pmatrix},$$

or, equivalently, by α-cut notation, for $\alpha \in [0; 1]$, we obtain

$$\tilde{A}_3 = \begin{pmatrix} [1;1] & [1+\frac{\alpha}{2}; 2-\frac{\alpha}{2}] & [7+\alpha; 9-\alpha] \\ [\frac{2}{4-\alpha}; \frac{2}{2+\alpha}] & [1;1] & [2+\alpha; 4-\alpha] \\ [\frac{1}{9-\alpha}; \frac{1}{7+\alpha}] & [\frac{2}{6-\alpha}; \frac{2}{4+\alpha}] & [1;1] \end{pmatrix}.$$

Here, problem (P1) has no feasible solution, hence, the weak consistency grade is $g(\tilde{A}_3) = \alpha^* = 0$, and therefore, $\tilde{A}_3$ is weak α-$\cdots$-inconsistent.

The priority vector of $\tilde{A}_3$ is obtained as the optimal solution of problem (P2), particularly, $w^{***} = (w_1^{***}, w_2^{***}, w_3^{***}) = (2.239, 1.201, 0.372)$.

The corresponding ranking of alternatives is $c_1 > c_2 > c_3$.

Inconsistency index $I.(\tilde{A}_3) = \mathscr{E}(A_3^*, w^*, 0) = 1.328 > 1$, hence, $\tilde{A}_3$ is $\cdots$ inconsistent.

It is evident that the 0-MPOP condition as well as the 0-MRP condition is met.

Example 4.12 Let $\tilde{A}_4 = \{\tilde{a}_{4ij}\}$ be a FPC matrix on the fuzzy multiplicative alo-group $\mathscr{R}_m = (]0, 1[, \bullet_f, \leq)$, with

$$a \bullet_f b = \frac{ab}{ab + (1-a)(1-b)}, e = 0.5, a^{(-1)} = 1 - a. \tag{4.61}$$

Fuzzy multiplicative alo-group $\mathscr{R}_m$ is divisible and continuous. For more details and properties (see Example 4.4, [12, 31]). Let $\tilde{A}_4$ be an FPC matrix with triangular fuzzy elements.

$$
\tilde{A}_4 = \begin{pmatrix} (0.5; 0.5; 0.5) & (0.45; 0.6; 0.7) & (0.4; 0.5; 0.65) \\ (0.3; 0.4; 0.55) & (0.5; 0.5; 0.5) & (0.3; 0.4; 0.55) \\ (0.35; 0.5; 0.6) & (0.45; 0.6; 0.7) & (0.5; 0.5; 0.5) \end{pmatrix},
$$

or, equivalently by α-cuts, $\alpha \in [0; 1]$, we obtain

$$
\tilde{A}_4 = \begin{pmatrix} [0.5; 0.5] & [0.45 + 0.15\alpha; 0.7 - 0.1\alpha] & [0.4 + 0.1\alpha; 0.65 - 0.15\alpha] \\ [0.3 + 0.1\alpha; 0.5 - 0.1\alpha] & [0.5; 0.5] & [0.3 + 0.1\alpha; 0.55 - 0.15\alpha] \\ [0.35 + 0.15\alpha; 0.6 - 0.1\alpha] & [0.45 + 0.15\alpha; 0.7 - 0.1\alpha] & [0.5; 0.5] \end{pmatrix}.
$$

Here, $\tilde{A}_4$ is a 3×3 FPC matrix, the elements of $\tilde{A}_4$ are triangular fuzzy numbers with the piecewise linear membership functions. Solving problem (P1), the $\bullet_f$-priority vector w^+ of $\tilde{A}_4$ is $w^+ = (0.533, 0.500, 0.467)$ and, the consistency grade $g_{\bullet_f}(\tilde{A}_4) = 0.556$. Hence, $\tilde{A}_4$ is α-$\bullet_f$-consistent for all $\alpha \leq 0.556$. Moreover, inconsistency index $I_{\bullet_f}(\tilde{A}_4) = 0.5$.

4.7 Pairwise Comparisons Matrices with Intuitionistic Fuzzy Elements

4.7.1 Introduction

Fuzzy sets being the elements of the pairwise comparison matrix (PCF matrix) may be applied whenever the decision-maker (DM) is not sure about the preference degree of his/her evaluation of the pairs in question. The intuitionistic fuzzy set (IFS), sometimes called Atanassov's IFS, is an extension of the fuzzy set, where the degree of non-membership denoting the non-belongingness to a set is explicitly specified along with the degree of membership of belongingness to the universal set, see [1]. Unlike the fuzzy set, where the non-membership degree is taken as one minus the membership degree, in IFS, the membership and non-membership degrees are more or less independent and related only by the condition that the sum of these two degrees must not exceed one.

Once again, we consider a decision-making problem (DM problem) which forms an application background in this chapter:

Let $\mathscr{C} = \{c_1, c_2, ..., c_n\}$ be a finite set of alternatives $(n > 2)$. The DM's aim is to rank the alternatives from the best to the worst (or, vice versa), using the information given by the DM in the form of an $n \times n$ FPC matrix.

The decision-maker acknowledges intuitionistic fuzzy pairwise preference data as imprecise knowledge about regular preference information. The preference matrix with *intuitionistic fuzzy* elements is then seen as a tool constraining an ill-known precise consistent comparison matrix. Inconsistencies, i.e., incompatibilities in comparison data are thus explicitly explained by the imprecise (or, inexact, vague, etc.) nature of human-originated information.

Usually, an ordinal ranking of alternatives is required to obtain the "best" alternative(s), however, it often occurs that the decision-maker is not satisfied with the ordinal ranking among alternatives and a cardinal ranking, i.e., *rating* is then required.

In the recent literature we can find papers dealing with applications of the pairwise comparison method where evaluations require fuzzy quantities or intuitionistic fuzzy quantities, for instance, when evaluating regional projects, web pages, e-commerce proposals, etc. (see e.g., [7, 19, 22, 26, 37]). In the papers [9, 13], Buckley et al. proposed a method for measuring elements of the pairwise comparisons matrix based on Saaty's principal eigenvector method. However, this method is rather cumbersome and numerically difficult. Moreover, it does not cover various uncertainties in the input data. The earliest work in AHP using fuzzy quantities as data was published by Van Laarhoven and Pedrycz [38]. They compared fuzzy ratios described by triangular membership functions. The method of logarithmic least squares (LLSQ) was used to derive local fuzzy priorities. Later on, using a geometric mean, Buckley [8] determined the fuzzy priorities of comparison ratios whose membership functions were assumed as trapezoidal. The issue of consistency in AHP using fuzzy sets as elements of the matrix was first tackled by Salo in [36]. Departing from the fuzzy arithmetic approach, fuzzy weights using an auxiliary mathematical programming formulation describing relative fuzzy ratios as constraints on the membership values of local priorities were derived. Later Leung and Cao [23] proposed a notion of tolerance deviation of fuzzy relative importance that is strongly related to Saaty's consistency ratio.

Earlier works that solved the problem of finding a rank of the given alternatives based on some FPC matrix are [23–28, 44]. In [44] some simple linear programming models for deriving the priority weights from various interval fuzzy preference relations are proposed. Leung and Cao [23] proposed a new definition of the FPC reciprocal matrix by setting deviation tolerances based on an idea of allowing inconsistent information. Mahmoudzadeh and Bafandeh [25] further discussed Leung and Cao's work and proposed a new method of fuzzy consistency test by direct fuzzification of a QR (Quick Response) algorithm, which is one of the numerical methods for calculating the eigenvalues of an arbitrary matrix. Ramik and Korviny in [33] investigated the inconsistency of a pairwise comparison matrix with fuzzy elements based on the geometric mean. They proposed an inconsistency index which, however, does not measure inconsistency as well as uncertainty ideally. In [31], the author presented a general approach for FPC matrices based on alo-groups which, in some sense, unifies the previous approaches. The recent paper is a continuation of this work extended to PC matrices with intuitionistic fuzzy intervals as the matrix entries.

Works on preference modeling and DM with intuitionistic fuzzy quantities can be found in numerous publications, particularly by Z. Xu and associates, see e.g., [42, 44–46], summarized later in the book [43]. Here, we generalized some approaches presented in these publications.

   This part of this chapter is organized as follows. In Sect. 4.7.2 we present some basic concepts and ideas, namely, the concept of the intuitionistic fuzzy set, and, in particular the bounded intuitionistic fuzzy interval of the abelian linearly ordered group (alo-group) of the real line. Section 4.7.3 is devoted to the pairwise comparison matrices with elements being fuzzy intervals (FPC matrices) and bounded intuitionistic fuzzy intervals (IFPC matrices). Reciprocity of IFPC matrices is introduced and some important basic properties are derived. Further on, the consistency of IFPC matrices is defined and some important basic properties are presented. Finally, the theory introduced initially for FPC matrices is extended to IFPC matrices and corresponding results are then derived including some methods for deriving priority vectors of IFPC matrices. The material and the results presented in Sect. 4.7.3, particularly in Sect. 4.3, are new and have not been published before. Some illustrative numerical examples are presented and discussed in order to clarify and explain the introduced concepts and results.

4.7.2 Preliminaries

Let us briefly recall the definition of fuzzy set as a special nested family of subsets of a universal set, introduced in Definition 4.1.

Definition 4.9 A *fuzzy subset* A of a nonempty set X (or a *fuzzy set A on X*) is a family $A = \{A_\alpha\}_{\alpha \in [0;1]}$ of subsets of X, such that $A_0 = X$, $A_\beta \subset A_\alpha$ whenever $0 \leq \alpha \leq \beta \leq 1$, and $A_\beta = \cap_{0 \leq \alpha < \beta} A_\alpha$ whenever $0 < \beta \leq 1$. The *membership function of A* is the function μ_A from X into the unit interval $[0; 1]$ defined by $\mu_A(x) = \sup\{\alpha \mid x \in A_\alpha\}$.

Similarly, an intuitionistic fuzzy (IF) set is a special couple of nested families of subsets of a set as follows.

Definition 4.10 An *intuitionistic fuzzy (IF) subset* C^I of a nonempty set X (or an *IF set on X*) is a couple of families $C^I = (A, B)$, $A = \{A_\alpha\}_{\alpha \in [0;1]}$ and $B = \{B_\alpha\}_{\alpha \in [0;1]}$, where A_α, B_α are subsets of X such that

$$A_0 = X, A_\beta \subset A_\alpha \ \text{whenever} \ 0 \leq \alpha \leq \beta \leq 1,$$
$$A_\beta = \cap_{0 \leq \alpha < \beta} A_\alpha \ \text{whenever} \ 0 < \beta \leq 1,$$
$$B_0 = X, B_\beta \subset B_\alpha \ \text{whenever} \ 0 \leq \alpha \leq \beta \leq 1,$$
$$B_\beta = \cap_{0 \leq \alpha < \beta} B_\alpha \ \text{whenever} \ 0 < \beta \leq 1,$$
$$A_\alpha \subset B_\alpha \ \text{whenever} \ 0 \leq \alpha \leq 1.$$

Each IF set C^I is characterized by a couple of functions: the *membership function of C^I* is the function μ_C from X into the unit interval $[0; 1]$ defined by

$$\mu_C(x) = \mu_A(x) = \sup\{\alpha \mid x \in A_\alpha\},$$

and the *non-membership function of* C^I is the function ν_C from X into the unit interval $[0; 1]$ defined by

$$\nu_C(x) = 1 - \mu_B(x), \text{ where } \mu_B(x) = \sup\{\alpha \mid x \in B_\alpha\}.$$

Let A be a subset of a set X and let $\{A_\alpha\}_{\alpha \in [0;1]}$ be the family of subsets of X defined by $A_0 = X$ and $A_\alpha = A$ for each positive α from $[0; 1]$ satisfying Definition 4.10. It can easily be seen that this family is a fuzzy set on X and that its membership function is equal to the characteristic function of A; we call it the *crisp fuzzy set* on X.

Remark 4.16 Each IF set $C^I = (A, B)$, where $A = \{A_\alpha\}$, $B = \{B_\alpha\}$, is given by two fuzzy sets. The first one, A, represents the membership, the other one, B, represents the non-membership of the IF set.

Remark 4.17 It is worth noting that an IF set $A^I = (A, A)$, where $A = \{A_\alpha\}$ is isomorphic to the fuzzy set A.

Moreover, each crisp set A is isomorphic to the IF fuzzy set $C^I = (A, A)$.

Remark 4.18 Notice that by the last inclusion in Definition 4.10, i.e., $A_\alpha \subset B_\alpha$ whenever $0 \le \alpha \le 1$, we obtain the standard well-known condition for IF sets (see Atanassov [1]),

$$\mu_A(x) + (1 - \mu_B(x)) \le 1 \text{ for all } x \in X.$$

We denote the collection of all fuzzy sets on X by $\mathscr{F}(X)$, similarly, the collection of all IF sets on X by $\mathscr{F}^I(X)$. For each $\alpha \in [0; 1]$, the set $[A]_\alpha = \{x \in X \mid \mu_A(x) \ge \alpha\}$ is called the α-*cut of fuzzy set* A. Similarly, for each $\alpha, \beta \in [0; 1]$, the set $\{x \in X \mid \mu_C(x) \ge \alpha, \nu_C(x) \le \beta\}$ is called the (α, β)-*cut of* IF set $C^I = (A, B)$ and it is denoted by $[C^I]_{\alpha,\beta}$. Notice that $[C^I]_{\alpha,\beta} = \{x \in X \mid \mu_A(x) \ge \alpha, \mu_B(x) \ge 1 - \beta\}$. If $\alpha = \beta$ we simply say that $[C^I]_{\alpha,\alpha}$ is the α-cut of IF set $C^I = (A, B)$ instead of the (α, α)-cut of the IF set. Likewise, we simply write $[C^I]_\alpha$ instead of $[C^I]_{\alpha,\alpha}$. Notice that

$$[C^I]_\alpha = \{x \in X \mid \mu_A(x) \ge \alpha, \mu_B(x) \ge 1 - \alpha\}. \tag{4.62}$$

If X is a nonempty subset of a real finite dimensional topological space, then a fuzzy set A in X is called *closed, bounded, compact*, or *convex* if the α-cut $[A]_\alpha$ is a closed, bounded, compact, or convex subset of X for every $\alpha \in]0; 1]$, respectively. Similarly, an IF set $C^I = (A, B)$ in X is called *closed, bounded, compact*, or *convex* if the (α, β)-cut $[C^I]_{\alpha,\beta}$ is a closed, bounded, compact, or convex subset of X for every $\alpha, \beta \in]0; 1[$, respectively.

In Sect. 4.2 we defined that a fuzzy subset A of $\mathbf{R}^* = \mathbf{R} \cup \{-\infty\} \cup \{+\infty\}$ is a *fuzzy interval* whenever A is normal and its membership function μ_A satisfies the following condition: there exist $a, b, c, d \in \mathbf{R}^*$, $-\infty \le a \le b \le c \le d \le +\infty$, such that

$\mu_A(t) = 0$ if $t < a$ or $t > d$,

$\quad\mu_A$ is strictly increasing and continuous on the interval $[a, b]$,

$\mu_A(t) = 1$ if $b \le t \le c$,   (4.63)

$\quad\mu_A$ is strictly decreasing and continuous on the interval $[c, d]$.

Moreover, we defined a *zero-cut of A* as $[A]_0 = [a, d]$. In Sect. 4.2 we introduced bounded, and compact intervals. The set of all bounded fuzzy intervals is denoted by $\mathscr{F}_0(\mathbf{R})$. The set of all fuzzy intervals is denoted by $\mathscr{F}_0(\mathbb{R}^*)$.

Remark 4.19 Notice that $\mathscr{F}_0(\mathbf{R})$ contains well-known classes of fuzzy numbers: crisp (real) numbers, crisp intervals, triangular fuzzy numbers, trapezoidal, and bell-shaped fuzzy numbers, etc. However, $\mathscr{F}_0(\mathbf{R})$ does not contain fuzzy sets with "stair-like" membership functions.

In a similar way, we extend the above stated concepts to IF sets. We say, that an IF set $C^I = (A, B)$ of $\mathbf{R}$ is an *IF interval*, resp. *bounded IF interval* whenever A and B are fuzzy intervals, resp. bounded fuzzy intervals. The set of all IF intervals is denoted by $\mathscr{F}_0^I(\mathbf{R}^*)$. The set of all bounded IF intervals is denoted by $\mathscr{F}_0^I(\mathbf{R})$.

Recall that the binary relations on X are subsets of the Cartesian product $X \times X$ and that the fuzzy sets on $X \times X$ are called the *fuzzy binary relation* on X, or simply the *fuzzy relation* on X. Because the set of the conventional binary relations on X can be embedded into the set of IF relations on X, we obtain an extension of conventional binary relation on X to the IF relation on X.

4.7.3 Pairwise Comparison Matrices with Elements Being Intuitionistic Fuzzy Intervals

Now, we shall investigate pairwise comparison matrices with elements being intuitionistic fuzzy intervals of the alo-group $\mathscr{G} = (G, \odot, \le)$ over an interval of the real line $\mathbf{R}$ (IFPC matrices). Alo-groups and their properties have been introduced in Chap. 4, Sect. 4.2. Such an approach allows for unifying the theory dealing with additive, multiplicative, and intuitionistic fuzzy—IFPC matrices (see e.g., [31]). In particular, we shall deal with IFPC matrices where the elements are intuitionistic fuzzy intervals. Moreover, we naturally assume that all diagonal elements of these matrices are crisp in the sense of Definition 4.9, and in particular they are equal to the identity element of $\mathscr{G}$, i.e. $\tilde{a}_{ii} = e$ for all $i \in \{1, \cdots, n\}$:

$$C^I = (\tilde{A}, \tilde{B}) = \begin{pmatrix} e & (\tilde{a}_{12}, \tilde{b}_{12}) & \cdots & (\tilde{a}_{1n}, \tilde{b}_{1n}) \\ (\tilde{a}_{21}, \tilde{b}_{21}) & e & \cdots & (\tilde{a}_{2n}, \tilde{b}_{2n}) \\ \vdots & \vdots & \ddots & \vdots \\ (\tilde{a}_{n1}, \tilde{b}_{n1}) & (\tilde{a}_{n2}, \tilde{b}_{n2}) & \cdots & e \end{pmatrix} \quad (4.64)$$

Here, $C^I = (\tilde{A}, \tilde{B})$ is an IF matrix with the elements $(\tilde{a}_{ij}, \tilde{b}_{ij})$, $i, j \in \{1, \cdots, n\}$, and $\tilde{a}_{ij}, \tilde{b}_{ij}$ are fuzzy intervals, particularly, fuzzy numbers. We shall call this matrix in short the *IFPC matrix*. For the sake of clarity, fuzzy intervals will be denoted by tilde above the corresponding symbol.

Let G be a proper interval of the real line $\mathbf{R}$ and $\le$ be the total order on G inherited from the usual order on $\mathbf{R}$, and $\mathscr{G} = (G, \odot, \le)$ be a real alo-group. We also assume that $\mathscr{G}$ is a divisible and continuous alo-group. Then, e.g., by [11], G is an open interval.

A *pairwise comparison system over* G is a pair $(\mathscr{C}, \tilde{A})$ constituted by a set $\mathscr{C} = \{c_1, c_2, \cdots, c_n\}$ and a mapping $\tilde{A} : \mathscr{C} \times \mathscr{C} \to \mathscr{F}_0^I(G)$, where $\tilde{A}(c_i, c_j) = (\tilde{a}_{ij}, \tilde{b}_{ij}) \in \mathscr{F}_0^I(G)$, and $\mathscr{F}_0^I(G)$ is the set of all bounded IF intervals and $C^I = \{(\tilde{a}_{ij}, \tilde{b}_{ij})\}$ is a IFPC matrix. In the context of a DM problem, the element $(\tilde{a}_{ij}, \tilde{b}_{ij})$ can be interpreted as an uncertain value on G of the DM's preference of c_i over c_j.

In most of the literature on fuzzy sets a fuzzy interval is defined by means of piecewise *linear* membership functions, but here we consider a more general case: strict monotone piecewise continuous functions. As IF intervals are understood as couples of fuzzy intervals, in the rest of this section we shall deal only with (bounded) fuzzy intervals.

Let $C^I = (\tilde{A}, \tilde{B})$ be an $n \times n$ IFPC matrix with the elements $(\tilde{a}_{ij}, \tilde{b}_{ij})$, $i, j \in \{1, \cdots, n\}$, and $\tilde{a}_{ij}, \tilde{b}_{ij}$ be fuzzy intervals with the membership function $\mu_{\tilde{a}_{ij}}$, $i, j \in \{1, \cdots, n\}$, $i \ne j$, (similarly, the membership function $\mu_{\tilde{b}_{ij}}$, $i, j \in \{1, \cdots, n\}$, $i \ne j$) be given as follows:

$$\mu_{\tilde{a}_{ij}}(x) = \begin{cases} L_{ij}^{-1}(x) & \text{if } x \in [L_{ij}(0), L_{ij}(1)], \\ 1 & \text{if } x \in [L_{ij}(1), R_{ij}(1)], \\ R_{ij}^{-1}(x) & \text{if } x \in [R_{ij}(1), R_{ij}(0)], \\ 0 & \text{otherwise}, \end{cases}$$

where L_{ij} is either an increasing continuous function, $L_{ij} : [0; 1] \to \mathbf{R}$ or eventually, L_{ij} is a constant function mapping interval $[0; 1]$ into a given point $y \in \mathbf{R}$. Also, R_{ij} is either a decreasing continuous function, $R_{ij} : [0; 1] \to \mathbf{R}$, or R_{ij} is a constant function mapping interval $[0; 1]$ into a given point $z \in \mathbf{R}$. We assume that $L_{ij}(1) \le R_{ij}(1)$. By L_{ij}^{-1}, or R_{ij}^{-1}, we denote the inverse functions of the increasing function L_{ij} or decreasing function R_{ij}, respectively. If L_{ij}, or R_{ij} is a constant function, then for $y \in \mathbf{R}$ we define $L_{ij}^{-1}(y) = R_{ij}^{-1}(y) = 1$.

Moreover, for all $i \in \{1, \cdots, n\}$ we assume that

$$\mu_{\tilde{a}_{ii}}(x) = \begin{cases} 1 & \text{if } x = e, \\ 0 & \text{otherwise}, \end{cases}$$

and

$$\mu_{\tilde{b}_{ii}}(x) = \begin{cases} 1 & \text{if } x = e, \\ 0 & \text{otherwise}, \end{cases}$$

where e is the *identity element of $\mathscr{G}$*. It is clear that each part of entry $\tilde{a}_{ij}$, or, $\tilde{b}_{ij}$, of the IFPC matrix $C^I = (\tilde{a}_{ij}, \tilde{b}_{ij})$ can be identified with a quadruple, e.g., $\tilde{a}_{ij} = (a_{ij}^L, a_{ij}^{LM}, a_{ij}^{RM}, a_{ij}^R)_{L_{ij}, R_{ij}}$, where $a_{ij}^L = L_{ij}(0)$, $a_{ij}^{LM} = L_{ij}(1)$, $a_{ij}^{RM} = R_{ij}(1)$, $a_{ij}^R = R_{ij}(0)$. For the sake of simplicity, if there is no danger of misunderstanding, we shall omit the subscripts referring to functions L_{ij}, R_{ij}, i.e., we simply write $\tilde{a}_{ij} = (a_{ij}^L, a_{ij}^{LM}, a_{ij}^{RM}, a_{ij}^R)$ and $\tilde{a}_{ij} = (a_{ij}^L, a_{ij}^{LM}, a_{ij}^{RM}, a_{ij}^R)$. The elements of the IFPC matrix with the above-mentioned properties will be called the *trapezoidal IF intervals*. In case $a_{ij}^{LM} = a_{ij}^{RM} = a_{ij}^M$, each element of the IFPC matrix will be called simply the *triangular IF interval*. Then they can be identified with triples, e.g., $\tilde{a}_{ij} = (a_{ij}^L, a_{ij}^M, a_{ij}^R)_{L_{ij}, R_{ij}}$, etc.

Remark 4.20 Notice that the crisp numbers (non-fuzzy numbers) are special cases of triangular fuzzy numbers and at the same time, special cases of bounded IF intervals. For instance, if for some $i, j \in \{1, \cdots, n\}$, $\tilde{a}_{ij}$ is a crisp number, then the corresponding functions L_{ij} and R_{ij} are constant and $\tilde{a}_{ij} = (a_{ij}^L, a_{ij}^M, a_{ij}^R) = (a_{ij}^L, a_{ij}^{LM}, a_{ij}^{RM}, a_{ij}^R)$ with $a_{ij}^L = a_{ij}^M = a_{ij}^{LM} = a_{ij}^{RM} = a_{ij}^R$. Notice that here the superscripts L, M, LM, RM and R mean "Left value", "Middle value", "Left Middle value", "Right Middle value", and "Right value", respectively.

Remark 4.21 The triangular IF numbers are also appropriate in group decision-making where, e.g., a^L can be interpreted as the minimum possible value of the DMs judgments, a^R is interpreted as the maximum possible value of their judgments, and a^M—the mean value of the DMs judgments—is interpreted as the mean value or the most possible value of their judgments (see e.g., [4]).

From now on, the following notation will be useful: Let $C^I = (\tilde{A}, \tilde{B}) = (\{\tilde{a}_{ij}\}, \{\tilde{b}_{ij}\})$ be an IFPC matrix $i, j \in \{1, \cdots, n\}$. Hence, the IFPC matrix C^I is given as a couple of matrices with elements being fuzzy intervals, called FPC matrices. Then we obtain the alpha-cuts as closed intervals

$$[\tilde{a}_{ij}]_\alpha = [a_{ij}^L(\alpha); a_{ij}^R(\alpha)], \quad [\tilde{b}_{ij}]_\alpha = [b_{ij}^L(\alpha); b_{ij}^R(\alpha)]. \tag{4.65}$$

Moreover, for all $\alpha \in\ ]0, 1]$, $i, j \in \{1, \cdots, n\}$ we have by Definition 4.10

$$[\tilde{a}_{ij}]_\alpha \subset [\tilde{b}_{ij}]_\alpha. \tag{4.66}$$

For $\alpha = 0$, we denote the 0-cuts as closed intervals

$$[\tilde{a}_{ij}]_0 = [a_{ij}^L(0); a_{ij}^R(0)], \quad [\tilde{b}_{ij}]_0 = [b_{ij}^L(0); b_{ij}^R(0)]. \tag{4.67}$$

4.7.4 IFPC Matrices, Reciprocity, and Consistency

Now, we extend the above stated definition of $\odot$-reciprocity and $\odot$-consistency to PC matrices with intuitionistic fuzzy elements, i.e., IFPCMs (see also [31]). In particular, we introduce new concepts of reciprocity and consistency based on α-cuts: α-$\odot$-reciprocity and α-$\odot$-consistency.

Definition 4.11 Let $\mathcal{G} = (G, \odot, \leq)$ be a divisible and continuous alo-group over an open interval G of $\mathbf{R}$. Let $\alpha \in [0; 1]$. Let $C^I = \{(\tilde{a}_{ij}, \tilde{b}_{ij})\}$ be an $n \times n$ matrix with intuitionistic elements, $\tilde{A} = \{\tilde{a}_{ij}\}$ and $\tilde{B} = \{\tilde{b}_{ij}\}$ be bounded fuzzy intervals of the alo-group $\mathcal{G}$. Let $[\tilde{a}_{ij}]_\alpha = [a^L_{ij}(\alpha); a^R_{ij}(\alpha)]$ and $[\tilde{b}_{ij}]_\alpha = [b^L_{ij}(\alpha); b^R_{ij}(\alpha)]$ be α-cuts.

Matrix $C^I = \{(\tilde{a}_{ij}, \tilde{b}_{ij})\}$ is said to be α-$\odot$-*reciprocal*, if the following conditions hold for each $i, j \in \{1, \cdots, n\}$:

$$a^L_{ii}(\alpha) = a^R_{ii}(\alpha) = b^L_{ii}(\alpha) = b^R_{ii}(\alpha) = e, \tag{4.68}$$

$$a^L_{ij}(\alpha) \odot a^R_{ji}(\alpha) = b^L_{ij}(\alpha) \odot b^R_{ji}(\alpha) = e. \tag{4.69}$$

If $C^I = \{(\tilde{a}_{ij}, \tilde{b}_{ij})\}$ is α-$\odot$-reciprocal for all $\alpha \in [0; 1]$, then it is called $\odot$-*reciprocal*.

If $C^I = \{(\tilde{a}_{ij}, \tilde{b}_{ij})\}$ is $\odot$-reciprocal, then $C^I = \{(\tilde{a}_{ij}, \tilde{b}_{ij})\}$ is called the *intuitionistic fuzzy pairwise comparisons matrix*, (or, *intuitionistic fuzzy PC matrix, IFPC matrix*, or, shortly, *IFPCM*).

Remark 4.22 Here, each element of the IFPC matrix is a couple of the corresponding elements of FPC matrices. The IFPC matrix is $\odot$-reciprocal if both FPC matrices are $\odot$-reciprocal in the sense of Sect. 4.3.

In practice, e.g., in AHP (see e.g., [35]), the crisp PC matrix is assumed to be reciprocal. Similarly, we ask that each IFPC matrix is $\odot$-reciprocal. In practice, this assumption is not too restrictive as the decision-maker usually evaluates only elements $(\tilde{a}_{ij}, \tilde{b}_{ij})$ for $1 \leq i < j \leq n$, and the elements $(\tilde{a}_{ji}, \tilde{b}_{ji})$ are given as the $\odot$-reciprocal ones.

Remark 4.23 In our approach, the membership functions of the intuitionistic fuzzy elements are not necessarily piecewise linear, which is a usual requirement in the literature, e.g., [18, 28, 39]. In our approach, only monotonicity of the membership function is required in the definition of bounded intuitionistic fuzzy interval. Such intuitionistic fuzzy elements may be either evaluated by individual decision-makers or they may be made up of crisp pairwise evaluations of decision-makers in a group DM problem, see e.g., [3].

Example 4.13 Consider $\odot = +$, let IFPC matrix $C^I = (\tilde{A}, \tilde{B}) = \{(\tilde{a}_{ij}, \tilde{b}_{ij})\}$ be as follows:

$$\tilde{A} = \begin{pmatrix} 0 & (1, 2, 4) & (4, 6, 7) \\ (-4, -2, -1) & 0 & (3, 4, 4) \\ (-7, -6, -4) & (-4, -4, -3) & 0 \end{pmatrix},$$

$$\tilde{B} = \begin{pmatrix} 0 & (1, 2, 5) & (4, 5, 8) \\ (-5, -2, -1) & 0 & (3, 4, 5) \\ (-8, -6, -4) & (-5, -4, -3) & 0 \end{pmatrix},$$

i.e.,

$$C^I = \begin{pmatrix} 0 & ((1,2,4),(1,2,5)) & ((4,6,7),(4,5,8)) \\ ((-4,-2,-1),(-5,-2,-1)) & 0 & ((3,4,4),(3,4,5)) \\ ((-7,-6,-4),(-8,-6,-4)) & ((-4,-4,-3),(-5,-4,-3)) & 0 \end{pmatrix},$$

Here, C^I is a 3×3 reciprocal IFPC matrix, specifically, an IFPC matrix with triangular IF number elements.

Now, we are going to extend the concept of α-$\odot$-consistency to IFPC matrices.

Definition 4.12 Let $\mathscr{G} = (G, \odot, \le)$ be a divisible and continuous alo-group over an open interval G of $\mathbf{R}$. Let $\alpha \in [0; 1]$. Let $C^I = \{(\tilde{a}_{ij}, \tilde{b}_{ij})\}$ be an $n \times n$ matrix with intuitionistic elements, $\tilde{A} = \{\tilde{a}_{ij}\}$ and $\tilde{B} = \{\tilde{b}_{ij}\}$ be bounded fuzzy intervals of the alo-group $\mathscr{G}$.

Matrix $C^I = \{(\tilde{a}_{ij}, \tilde{b}_{ij})\}$ is said to be *weak α-$\odot$-consistent*, if the both the bounded fuzzy intervals $\tilde{A} = \{\tilde{a}_{ij}\}$ and $\tilde{B} = \{\tilde{b}_{ij}\}$ are weak α-$\odot$-consistent by Definition 4.2. Moreover, if $C^I = \{(\tilde{a}_{ij}, \tilde{b}_{ij})\}$ is weak α-$\odot$-consistent for all $\alpha \in [0; 1]$, then it is *weak $\odot$-consistent*.

If for some $\alpha \in [0; 1]$ the IFPC matrix $C^I = \{(\tilde{a}_{ij}, \tilde{b}_{ij})\}$ is not weak α-$\odot$-consistent, then C^I is called $\odot$-*inconsistent*.

Remark 4.24 In other words, Definition 4.12 says that each IFPC matrix $C^I = \{(\tilde{a}_{ij}, \tilde{b}_{ij})\}$, if the both the bounded fuzzy intervals $\tilde{A} = \{\tilde{a}_{ij}\}$ and $\tilde{B} = \{\tilde{b}_{ij}\}$ are weak α-$\odot$-consistent. Hence, the results derived in Sect. 4.2 for weak $\odot$-consistent FPC matrices can be applied to IFPC matrices, e.g., Proposition 4.2. In particular, for all $\alpha \in]0, 1]$, $i, j \in \{1, \cdots, n\}$ we have by Definition 4.10

$$[\tilde{a}_{ij}]_\alpha \subset [\tilde{b}_{ij}]_\alpha, \tag{4.70}$$

hence, IFPC matrix $C^I = \{(\tilde{a}_{ij}, \tilde{b}_{ij})\}$ is weak α-$\odot$-consistent, if and only if the first part of C^I, i.e., FPCM $\tilde{A} = \{\tilde{a}_{ij}\}$, is weak α-$\odot$-consistent.

Now, we are going to extend the stronger concept of α-$\odot$-consistency to IFPC matrices applying Definition 4.4 and Proposition 4.2, (iii).

Definition 4.13 Let $\mathscr{G} = (G, \odot, \le)$ be a divisible and continuous alo-group over an open interval G of $\mathbf{R}$. Let $\alpha \in [0; 1]$. Let $C^I = \{(\tilde{a}_{ij}, \tilde{b}_{ij})\}$ be an $n \times n$ matrix with intuitionistic elements, $\tilde{A} = \{\tilde{a}_{ij}\}$ and $\tilde{B} = \{\tilde{b}_{ij}\}$ be bounded fuzzy intervals of the alo-group $\mathscr{G}$.

Matrix $C^I = \{(\tilde{a}_{ij}, \tilde{b}_{ij})\}$ is said to be α-$\odot$-*consistent*, if the both the bounded fuzzy intervals $\tilde{A} = \{\tilde{a}_{ij}\}$ and $\tilde{B} = \{\tilde{b}_{ij}\}$ are α-$\odot$-consistent by Definition 4.4. Moreover, if $C^I = \{(\tilde{a}_{ij}, \tilde{b}_{ij})\}$ is α-$\odot$-consistent for all $\alpha \in [0; 1]$, then it is $\odot$-*consistent*.

We already know that an α-$\odot$-consistent FPC matrix is not necessarily α-$\odot$-reciprocal. In Definition 4.13, α-$\odot$-reciprocity of the IFPC matrix is not assumed.

In real DM problems, the α-$\odot$-reciprocity condition is, however, a natural assumption. The following proposition gives a characterization of α-$\odot$-reciprocal α-$\odot$-consistent IFPC matrices (see also [10] or [24, 39]). The proof follows directly from Proposition 4.2.

Proposition 4.10 *Let $\alpha \in [0; 1]$, let $C^I = \{(\tilde{a}_{ij}, \tilde{b}_{ij})\}$ be an IFPC matrix, $\tilde{A} = \{\tilde{a}_{ij}\}$ and $\tilde{B} = \{\tilde{b}_{ij}\}$ be α-$\odot$-reciprocal bounded fuzzy intervals of the alo-group $\mathscr{G}$. Let $[\tilde{a}_{ij}]_\alpha = [a_{ij}^L(\alpha); a_{ij}^R(\alpha)]$ and $[\tilde{b}_{ij}]_\alpha = [b_{ij}^L(\alpha); b_{ij}^R(\alpha)]$ be α-cuts. The following conditions are equivalent.*

(i) $C^I = \{(\tilde{a}_{ij}, \tilde{b}_{ij})\}$ *is α-$\odot$-consistent.*
(ii) *For all $i, j, k \in \{1, \cdots, n\}$, it holds*

$$a_{ik}^L(\alpha) \odot a_{ik}^R(\alpha) = a_{ij}^L(\alpha) \odot a_{ij}^R(\alpha) \odot a_{jk}^L(\alpha) \odot a_{jk}^R(\alpha), \qquad (4.71)$$

and

$$b_{ik}^L(\alpha) \odot b_{ik}^R(\alpha) = b_{ij}^L(\alpha) \odot b_{ij}^R(\alpha) \odot b_{jk}^L(\alpha) \odot b_{jk}^R(\alpha). \qquad (4.72)$$

Definition 4.14 Let $\alpha \in [0; 1]$, let $C^I = \{(\tilde{a}_{ij}, \tilde{b}_{ij})\}$ be a IFPC matrix, $[\tilde{a}_{ij}]_\alpha = [a_{ij}^L(\alpha), a_{ij}^R(\alpha)]$, $[\tilde{b}_{ij}]_\alpha = [b_{ij}^L(\alpha), b_{ij}^R(\alpha)]$ be corresponding α-cuts. For all $i, j \in \{1, ..., n\}$ denote

$$a_{ij}^m(\alpha) = (a_{ij}^L(\alpha) \odot a_{ij}^R(\alpha))^{(\frac{1}{2})}. \qquad (4.73)$$

$$b_{ij}^m(\alpha) = (b_{ij}^L(\alpha) \odot b_{ij}^R(\alpha))^{(\frac{1}{2})}. \qquad (4.74)$$

A crisp $n \times n$-matrix $A^m(\alpha) = \{a_{ij}^m(\alpha)\}$ is called the α-$\odot$-*mean matrix* associated with FPC matrix $\tilde{A} = \{\tilde{a}_{ij}\}$, matrix $B^m(\alpha) = \{b_{ij}^m(\alpha)\}$ is called the α-$\odot$-*mean matrix* associated with FPC matrix $\tilde{B} = \{\tilde{b}_{ij}\}$.

By Proposition 4.2, (iii), an IFPC matrix $C^I = \{(\tilde{a}_{ij}, \tilde{b}_{ij})\}$ is α-$\odot$-consistent, if and only if the both crisp α-$\odot$-mean matrix $A^m(\alpha)$ and $B^m(\alpha)$ are $\odot$-consistent, i.e., the following formulas hold for all $i, j, k \in \{1, \cdots, n\}$:

$$a_{ik}^m(\alpha) = a_{ij}^m(\alpha) \odot a_{jk}^m(\alpha). \qquad (4.75)$$

$$b_{ik}^m(\alpha) = b_{ij}^m(\alpha) \odot b_{jk}^m(\alpha). \qquad (4.76)$$

Remark 4.25 Notice that the α-$\odot$-consistency of IFPC matrix $C^I = \{(\tilde{a}_{ij}, \tilde{b}_{ij})\}$ is based on $\odot$-consistency of the associated two crisp matrices, in particular, α-$\odot$-mean matrices associated with $\tilde{A}$ and $\tilde{B}$, with elements being generalized means (4.23) of the end points of the corresponding α-cuts. This fact will be advantageous in deriving a corresponding priority vector of the IFPC matrix as we can see in the next subsection.

Example 4.14 Consider the additive alo-group $\mathscr{R} = (\mathbf{R}, \odot, \leq)$ with $\odot = +$ (see Example 4.1). Let $C^I = \{(\tilde{A}, \tilde{B})\}$, where $\tilde{A} = \{\tilde{a}_{ij}\}$, $\tilde{B} = \{\tilde{b}_{ij}\}$, be given as follows:

$$\tilde{A} = \begin{pmatrix} (0,0,0) & (5,6,7) & (12,14,16) \\ (-7,-6,-5) & (0,0,0) & (7,8,9) \\ (-16,-14,-11) & (-9,-8,-6) & (0,0,0) \end{pmatrix},$$

$$\tilde{B} = \begin{pmatrix} (0,0,0) & (4,6,8) & (10,14,18) \\ (-8,-6,-4) & (0,0,0) & (6,8,10) \\ (-18,-14,-10) & (-10,-8,-6) & (0,0,0) \end{pmatrix},$$

or, equivalently, by the definition of fuzzy sets in Sect. 4.2, each fuzzy set is given by the corresponding family of α-cuts. In particular, for $\alpha \in [0; 1]$, we obtain

$$\tilde{A} = \begin{pmatrix} [0;0] & [5+\alpha; 7-\alpha] & [12+2\alpha; 16-2\alpha] \\ [-7+\alpha; -5-\alpha] & [0;0] & [6+2\alpha; 10-2\alpha] \\ [-16+2\alpha; -11-3\alpha] & [-10+2\alpha; -6-2\alpha] & [0;0] \end{pmatrix},$$

$$\tilde{B} = \begin{pmatrix} [0;0] & [4+2\alpha; 8-2\alpha] & [10+4\alpha; 18-4\alpha] \\ [-8+2\alpha; -4-2\alpha] & [0;0] & [6+2\alpha; 10-2\alpha] \\ [-18+4\alpha; -10-4\alpha] & [-10+2\alpha; -6-2\alpha] & [0;0] \end{pmatrix}.$$

Evidently, $\tilde{A} \subset \tilde{B}$, hence, C^I is an IFPC matrix with the corresponding piecewise linear membership functions.

Moreover, α-+-mean matrix $A^m(\alpha)$ and $B^m(\alpha)$ associated to $\tilde{A}$ and $\tilde{B}$, respectively, for $\alpha \in [0; 1]$ is calculated by (4.23) as

$$A^m(\alpha) = \begin{pmatrix} 0 & 6 & 14 \\ -6 & 0 & 8 \\ -14 & -8 & 0 \end{pmatrix},$$

$$B^m(\alpha) = \begin{pmatrix} 0 & 6 & 14 \\ -6 & 0 & 8 \\ -14 & -8 & 0 \end{pmatrix}.$$

We obtained $A^m(\alpha) = B^m(\alpha)$ and, moreover, by checking equality (4.75), (4.76), we obtain that $\tilde{A}$ and $\tilde{B}$ are α-+-consistent FPCMs for all $0 \leq \alpha \leq 1$, i.e., they are +-consistent. Notice that both $A^m(\alpha)$, and $B^m(\alpha)$ are independent of α. Therefore, $C^I = \{(\tilde{A}, \tilde{B})\}$ is a +-consistent IFPC matrix.

The following propositions give some characterizations of α-$\odot$-consistent IFPC matrices. The proof follows directly from Proposition 3.3.

Proposition 4.11 *Let* $\alpha \in [0; 1]$, *let* $C^I = \{(\tilde{a}_{ij}, \tilde{b}_{ij})\}$ *be an IFPC matrix,* $[\tilde{a}_{ij}]_\alpha = [a_{ij}^L(\alpha); a_{ij}^R(\alpha)]$ *and* $[\tilde{b}_{ij}]_\alpha = [b_{ij}^L(\alpha); b_{ij}^R(\alpha)]$ *be* α-cuts. *The following conditions are equivalent.*

(i) $C^I = \{(\tilde{a}_{ij}, \tilde{b}_{ij})\}$ is α-$\odot$-consistent.
(ii) There exist vectors $v(\alpha) = (v_1(\alpha), \cdots, v_n(\alpha))$, $w(\alpha) = (w_1(\alpha), \cdots, w_n(\alpha))$ with $v_j(\alpha), w_j(\alpha) \in G$, $j \in \{1, \cdots, n\}$, such that for each $i, k \in \{1, \cdots, n\}$, it holds that

$$a_{ik}^m(\alpha) = v_i(\alpha) \div v_k(\alpha). \tag{4.77}$$

$$b_{ik}^m(\alpha) = w_i(\alpha) \div w_k(\alpha). \tag{4.78}$$

From Proposition 4.4 we easily obtain the following property:

Proposition 4.12 Let $\alpha \in [0; 1]$, let $C^I = \{(\tilde{a}_{ij}, \tilde{b}_{ij})\}$ be an IFPC matrix.
If $C^I = \{(\tilde{a}_{ij}, \tilde{b}_{ij})\}$ is α-$\odot$-consistent then $C^I = \{(\tilde{a}_{ij}, \tilde{b}_{ij})\}$ is weak α-$\odot$-consistent.
Moreover, if C^I is $\odot$-consistent then C^I is weak $\odot$-consistent.

Example 4.15 Consider the additive alo-group $\mathcal{R} = (\mathbf{R}, \odot, \leq)$ with $\odot = +$ (see Example 4.1), $\alpha \in [0; 1]$. Let $C^I = \{(\tilde{A}, \tilde{B})\}$, where $\tilde{A} = \{\tilde{a}_{ij}\}$, $\tilde{B} = \{\tilde{b}_{ij}\}$, be given as follows:

$$\tilde{A} = \begin{pmatrix} (0,0,0) & (1,3,4) & (4,6,9) \\ (-4,-3,-1) & (0,0,0) & (2,4,5) \\ (-9,-6,-4) & (-5,-4,-2) & (0,0,0) \end{pmatrix},$$

$$\tilde{B} = \begin{pmatrix} (0,0,0) & (1,3,5) & (2,6,11) \\ (-5,-3,-1) & (0,0,0) & (1,4,6) \\ (-11,-6,-2) & (-6,-4,-1) & (0,0,0) \end{pmatrix},$$

Evidently, $\tilde{A} \subset \tilde{B}$, $\tilde{A}$ and $\tilde{B}$ are FPC matrices with triangular fuzzy number elements and the corresponding piecewise linear membership functions.
We obtain that $\tilde{A}$ is weak α-+-consistent for all $0 \leq \alpha \leq \frac{6}{7}$ and also $\tilde{B}$ is weak α-+-consistent for all $0 \leq \alpha \leq \frac{9}{10}$.
Therefore, $C^I = \{(\tilde{A}, \tilde{B})\}$ is *weak* α-+-consistent for all $0 \leq \alpha \leq \min\{\frac{6}{7}, \frac{9}{10}\} = \frac{6}{7}$.

4.7.5 Priority Vectors of IFPC Matrices

Here, we propose a method for calculating the priority vector of $n \times n$ IFPC matrix $C^I = \{(\tilde{a}_{ij}, \tilde{b}_{ij})\}$ for the purpose of rating the alternatives $c_1, \cdots, c_n \in \mathscr{C}$. As it has been stated before, we do not follow the way of calculating the *fuzzy* priority vector proposed, e.g., in [13, 15] and others. Here, we generate a crisp priority vector, therefore, so no defuzzification is necessary for final ranking of the alternatives.
Similarly to Sect. 4.5, the proposed method for calculating the priority vector can be divided into two steps as follows.

Step 1.

In *Step 1* we check whether the initial IFPC matrix $C^I = \{(\tilde{a}_{ij}, \tilde{b}_{ij})\}$ is weak α-$\odot$-consistent for some α, where $0 \leq \alpha \leq 1$. We calculate the maximal such α denoted by α^*. Therefore, IFPCM C^I is weak α-$\odot$-consistent for all α, $0 \leq \alpha \leq \alpha^*$. By Proposition 4.11, (ii), the following optimization problem is to be solved:

(IP1)

$$\alpha \longrightarrow \max; \tag{4.79}$$

subject to

$$a_{ij}^L(\alpha) \leq w_i \div w_j \leq a_{ij}^R(\alpha) \text{ for all } i, j \in \{1, \cdots, n\}, \tag{4.80}$$

$$\bigodot_{k=1}^{n} w_k = e, \tag{4.81}$$

$$0 \leq \alpha \leq 1, w_k \in G, \text{ for all } k \in \{1, \cdots, n\}. \tag{4.82}$$

In problem (IP1), the objective function (4.79) is maximized under the constraints securing that that FPC matrix $\{\tilde{a}_{ij}\}$ is weak α-$\odot$-consistent.

As $\tilde{a}_{ij} \subset \tilde{b}_{ij}$, for all $i, j \in \{1, \cdots, n\}$, the feasible solution of (IP1), i.e., vector $w = (w_1, \cdots, w_n)$, satisfies also

$$b_{ij}^L(\alpha) \leq w_i \div w_j \leq b_{ij}^R(\alpha) \text{ for all } i, j \in \{1, \cdots, n\}. \tag{4.83}$$

Therefore, by Proposition 4.1, FPCM $\{\tilde{b}_{ij}\}$ is also weak α-$\odot$-consistent. Consequently, IFPC matrix $C^I = \{(\tilde{a}_{ij}, \tilde{b}_{ij})\}$ is weak α-$\odot$-consistent and $w = (w_1, \cdots, w_n)$ in (4.81) is normalized.

If optimization problem (IP1) has a feasible solution, i.e., the system of constraints (4.80)–(4.82) has a solution, then it is clear that (IP1) also has an optimal solution.

Let α^* and $w^1 = (w_1^1, \cdots, w_n^1)$ be an optimal solution of problem (IP1). Then α^* is called the *weak $\odot$-consistency grade of IFPC matrix C^I*, denoted by $g_\odot(C^I)$, i.e., we define

$$g_\odot(C^I) = \alpha^*. \tag{4.84}$$

Here, $0 \leq \alpha^* \leq 1$. Moreover, if $g_\odot(C^I) = 1$, then IFPC matrix C^I is weak $\odot$-consistent.

If optimization problem (IP1) has **no** feasible solution, which means that IFPC matrix $C^I = \{(\tilde{a}_{ij}, \tilde{b}_{ij})\}$ is weak α-$\odot$-consistent for no $\alpha \in [0; 1]$, then we define

$$g_\odot(C^I) = 0. \tag{4.85}$$

In that case, the corresponding priority vector will be defined in what follows.

Go to Step 2.

Remark 4.26 In general, problem (IP1) is a nonlinear optimization problem that may be solved by some appropriate numerical methods, e.g., by the well-known dichotomy method, which is a sequence of relatively simple optimization problems, see e.g., [29], or by some gradient-type method (see [6]).

In the next step we obtain a corresponding priority vector, eventually, with our desirable properties.

Step 2.
First, assume that problem (IP1) is feasible. Here, the elements of crisp PC matrix $A^* = \{a_{ij}^*\}$ are defined by (4.23) as

$$a_{ij}^* = (a_{ij}^L(\alpha^*) \odot a_{ij}^R(\alpha^*))^{(\frac{1}{2})},$$

and of crisp PC matrix $B^* = \{b_{ij}^*\}$ are defined by (4.31) as

$$b_{ij}^* = (b_{ij}^L(\alpha^*) \odot b_{ij}^R(\alpha^*))^{(\frac{1}{2})},$$

where $\alpha^* \in [0; 1]$ has been calculated in Step 1, $\alpha^* = g_{\odot}(C')$.
Let $A^* = \{a_{ij}^*\} \in PC_n(\mathcal{G})$ be a PC matrix, $\varepsilon > e$ be a suitable constant. Define the following two sets of indexes:

$$I^{(1)}(A^*) = \{(i, j) | i, j \in \{1, \cdots, n\}, a_{ij}^* = e\}, \tag{4.86}$$

$$I^{(2)}(A^*) = \{(i, j) | i, j \in \{1, \cdots, n\}, a_{ij}^* > e\}, \tag{4.87}$$

An *error index*, $\mathscr{E}(A^*, w, , \alpha^*)$, of $A^* = \{a_{ij}^*\}$ and $w = (w_1, \cdots, w_n)$ has been defined by (4.35) as

$$\mathscr{E}(A^*, w, \alpha^*) = \max\{\|a_{ij}^* \odot w_j \div w_i\| | i, j \in \{1, \cdots, n\}\}. \tag{4.88}$$

Now, we solve problem (IP2) as follows.

(IP2)

$$\mathscr{E}(A^*, w, \alpha^*) \longrightarrow \min; \tag{4.89}$$

subject to

$$a_{ij}^L(\alpha^*) \le w_i \div w_j \le a_{ij}^R(\alpha^*) \text{ for all } i, j \in \{1, \cdots, n\}, \tag{4.90}$$

$$\bigodot_{k=1}^{n} w_k = e, \quad w_k \in G, \quad k \in \{1, \cdots, n\}, \tag{4.91}$$

and, eventually, subject to the desirable properties FS, POP, and RP

$$w_r = w_s \ \forall (r, s) \in I^{(1)}(A^*), \tag{4.92}$$

$$w_r \geq w_s \odot \varepsilon \quad \forall (r, s) \in I^{(2)}(A^*), \tag{4.93}$$

The optimal solution $w^* = (w_1^*, \cdots, w_n^*)$ of (IP2) will be called the $\odot$-*priority vector of* C^I.

Problem (IP2) with variables $w_1, \cdots, w_n$ is feasible, as $w^1 = (w_1^1, \cdots, w_n^1)$ is a part of a feasible solution of problem (IP1). Hence, it is also a feasible solution of (IP2) with objective function (4.55).

Second, assume that problem (IP1) is *infeasible*. By solving problem (IP2), now without constraints (4.56), with $\alpha^* = 0$, where α^*-$\odot$-mean matrix $A^* = \{a_{ij}^*\}$ is associated with FPC matrix $\tilde{A} = \{\tilde{a}_{ij}\}$, see (4.23), we obtain a corresponding priority vector $w^* = (w_1^*, \cdots, w_n^*)$, eventually with desirable properties FS, POP, and RP, i.e., (4.94), (4.95).

Here, however, the elements of crisp PC matrix $A^* = \{a_{ij}^*\}$ are defined by (4.23) as

$$a_{ij}^* = (a_{ij}^L(0) \odot a_{ij}^R(0))^{(\frac{1}{2})}.$$

In order to obtain an $\odot$-priority vector satisfying desirable properties FS, POP, and RP, we have to solve problem (IP2) with two additional constraints

$$w_r = w_s \quad \forall (r, s) \in I^{(1)}(A^*), \tag{4.94}$$

$$w_r \geq w_s \odot \varepsilon \quad \forall (r, s) \in I^{(2)}(A^*). \tag{4.95}$$

The existence of such optimal solution satisfying properties (4.94) and (4.95) is, however, not secured.

Remark 4.27 In general, the uniqueness of the optimal solution of (IP2) is not preserved. Depending on the particular operation $\odot$, problem (IP2) may have multiple optimal solutions which is an unfavorable property from the point of view of the DM. In this case, the DM should reconsider some (fuzzy) evaluations in the original pairwise comparison matrix. Consequently, we obtain the following proposition.

4.7.6 Measuring Inconsistency of IFPC Matrices

If for some $\alpha \in [0; 1]$ the IFPC matrix $C^I = \{(\tilde{A}, \tilde{B})\}$ is not weak α-$\odot$-consistent, then C^I is considered to be α-$\odot$-*inconsistent*. Now, we introduce a concept of measuring inconsistency by an inconsistency index.

Let $C^I = \{(\tilde{A}, \tilde{B})\}$ be an IFPCM and let $A = \{a_{ij}\}$ denote a PCM with the elements from the 0-cut of $\tilde{A} = \{\tilde{a}_{ij}\}$, i.e., $a_{ij} \in [\tilde{a}_{ij}]_0, i, j \in \{1, \cdots, n\}$. Similarly, by $B = \{b_{ij}\}$ denote a PCM with the elements from the 0-cut of $\tilde{B} = \{\tilde{b}_{ij}\}$, i.e., $b_{ij} \in [\tilde{b}_{ij}]_0, i, j \in \{1, \cdots, n\}$.

The $\odot$-*inconsistency index of* $C^I = \{(\tilde{A}, \tilde{B})\}$, $I_\odot(C^I)$ is defined as

$$I_\odot(C^I) = \max\{I_\odot(\tilde{A}), I_\odot(\tilde{B})\}, \tag{4.96}$$

where

$$I_\odot(\tilde{A}) = \inf\{\sup\{\mathscr{E}(A, v, 0)|a_{ij} \in [\tilde{a}_{ij}]_0, i, j \in \{1, \cdots, n\}\}| \bigodot_{k=1}^{n} v_k = e\}. \tag{4.97}$$

and

$$I_\odot(\tilde{B}) = \inf\{\sup\{\mathscr{E}(B, w, 0)|b_{ij} \in [\tilde{b}_{ij}]_0, i, j \in \{1, \cdots, n\}\}| \bigodot_{k=1}^{n} w_k = e\}. \tag{4.98}$$

Remark 4.28 As $\tilde{a}_{ij} \subset \tilde{b}_{ij}$, for all $i, j \in \{1, \cdots, n\}$, we easily obtain

$$I_\odot(\tilde{A}) \leq I_\odot \tilde{B}). \tag{4.99}$$

Hence,

$$I_\odot(C^I) = I_\odot(\tilde{B}). \tag{4.100}$$

Consequently, we obtain the following proposition.

Proposition 4.13 *If $C^I = \{(\tilde{a}_{ij}, \tilde{b}_{ij})\}$ is an IFPC matrix, then exactly one of the following two cases occurs:*

(i) *Problem (IP1) has a feasible solution α^*. Then weak consistency grade $g_\odot(C^I) = \alpha^*$, $0 \leq \alpha^* \leq 1$.*
   For each α, such that $0 \leq \alpha \leq \alpha^* \leq 1$, IFPC matrix $C^I = \{(\tilde{a}_{ij}, \tilde{b}_{ij})\}$ is weak α-$\odot$-consistent. The associated priority vector $w^* = (w_1^*, \cdots, w_n^*)$ is the optimal solution of (IP2).
(ii) *Problem (IP1) has no feasible solution. Then consistency grade $g_\odot(C^I) = 0$, C^I is $\odot$-inconsistent, hence $I_\odot(C^I) > e$. The associated priority vector $w^* = (w_1^*, \cdots, w_n^*)$ is the optimal solution of (IP2).*

Remark 4.29 Inconsistency index (4.100) and (4.98) may be calculated by a suitable "min-max" optimization method (see e.g., [6]).

Remark 4.30 Evidently, if C^I is a crisp IFPC matrix, then $\odot$-inconsistency index $I_\odot(C^I) = e$, if and only if A is (weak) $\odot$-consistent.

Example 4.16 Consider the additive alo-group $\mathscr{R} = (\mathbf{R}, +, \leq)$ with $\odot = +$, see Example 4.1. Let for the three alternatives $\mathscr{C} = \{c_1, c_2, c_3\}$, the IFPCM $C^I = \{(\tilde{a}_{ij}, \tilde{b}_{ij})\}$, $\tilde{A} = \{\tilde{a}_{ij}\}$, $\tilde{\mathscr{B}} = \{\tilde{b}_{ij}\}$ be given by triangular fuzzy number elements as follows:

$$\tilde{A} = \begin{pmatrix} (0, 0, 0) & (5, 6, 7) & (1, 2, 3) \\ (-7, -6, -5) & (0, 0, 0) & (-5, -2, -1) \\ (-3, -2, -1) & (1, 2, 5) & (0, 0, 0) \end{pmatrix},$$

or, equivalently, by α-cut notation, for $\alpha \in [0; 1]$, we obtain

$$\tilde{A} = \begin{pmatrix} [0;0] & [5+\alpha; 7-\alpha] & [1+\alpha; 3-\alpha] \\ [-7+\alpha; -5-\alpha] & [0;0] & [-5+3\alpha; -1-\alpha] \\ [-3+\alpha; -1-\alpha] & [1+\alpha; 5-3\alpha] & [0;0] \end{pmatrix}.$$

$$\tilde{B} = \begin{pmatrix} (0,0,0) & (4,6,8) & (1,2,4) \\ (-8,-6,-4) & (0,0,0) & (-6,-2,-1) \\ (-4,-2,-1) & (1,2,6) & (0,0,0) \end{pmatrix},$$

or, equivalently, by α-cut notation, for $\alpha \in [0; 1]$, we obtain

$$\tilde{B} = \begin{pmatrix} [0;0] & [4+2\alpha; 8-2\alpha] & [1+\alpha; 4-2\alpha] \\ [-8+2\alpha; -4-2\alpha] & [0;0] & [-6+4\alpha; -1-\alpha] \\ [-4+2\alpha; -1-\alpha] & [1+\alpha; 6-4\alpha] & [0;0] \end{pmatrix}.$$

Here, $\tilde{B}$ is an FPC matrix with triangular fuzzy number elements. Solving problem (IP1), we obtain the weak consistency grade as

$$g(\tilde{A}) = \alpha^* = 0.333,$$

specifically, C^I is weak α-$\odot$-consistent for all $0 \le \alpha \le 0.333$.

The priority vector is obtained as the optimal solution of problem (IP2), specifically,

$$w^* = (w_1^*, w_2^*, w_3^*) = (2.667, -2.889, 0.222).$$

The corresponding ranking of alternatives is $c_1 \succ c_3 \succ c_2$.

Example 4.17 Consider once again the additive alo-group $\mathscr{R} = (\mathbf{R}, +, \le)$ with $\odot = +$ (see Example 4.1). Let for the three alternatives $\mathscr{C} = \{c_1, c_2, c_3\}$, IFPCM $C^I = \{(\tilde{a}_{ij}, \tilde{b}_{ij})\}$, with $\tilde{A} = \{\tilde{a}_{ij}\}$, $\tilde{B} = \{\tilde{b}_{ij}\}$ be given by triangular fuzzy number elements as:

$$\tilde{A} = \begin{pmatrix} (0,0,0) & (1,2,2.5) & (7,8,9) \\ (-2.5,-2,-1) & (0,0,0) & (2,3,3.5) \\ (-9,-8,-7) & (-3.5,-3,-2) & (0,0,0) \end{pmatrix},$$

or, equivalently, by α-cut notation, for $\alpha \in [0; 1]$, we obtain

$$\tilde{A} = \begin{pmatrix} [0;0] & [1+\alpha; 2.5-0.5\alpha] & [7+\alpha; 9-\alpha] \\ [-2.5+0.5\alpha; -1-\alpha] & [0;0] & [2+\alpha; 3.5-0.5\alpha] \\ [-9+\alpha; -7-\alpha] & [-3.5+0.5\alpha; -2-\alpha] & [0;0] \end{pmatrix}.$$

$$\tilde{B} = \begin{pmatrix} (0,0,0) & (0.5,2,3) & (6,8,10) \\ (-3,-2,-0.5) & (0,0,0) & (1,3,4) \\ (-10,-8,-6) & (-4,-3,-1) & (0,0,0) \end{pmatrix},$$

or, equivalently, by α-cut notation, for $\alpha \in [0; 1]$, we obtain

$$\tilde{B} = \begin{pmatrix} [0; 0] & [0.5 + 1.5\alpha; 2 + \alpha] & [6 + 2\alpha; 10 - 2\alpha] \\ [-2 - \alpha; -0.5 - 1.5\alpha] & [0; 0] & [1 + 2\alpha; 4 - \alpha] \\ [-10 + 2\alpha; -6 - 2\alpha] & [-4 + \alpha; -1 - 2\alpha] & [0; 0] \end{pmatrix}.$$

Problem (IP1) is infeasible, hence the weak consistency grade is

$$g(\tilde{A}) = \alpha^+ = 0.000.$$

Therefore, C^I is weak α-$\odot$-consistent for no $0 \leq \alpha \leq 1$.

The priority vector is obtained as the optimal solution of problem (IP2), specifically,

$$w^+ = (w_1^+ w_2^+, w_3^+) = (4.167, 0.333, -4.500).$$

The corresponding ranking of alternatives is $c_1 \succ c_2 \succ c_3$.

Inconsistency index is calculated by (4.100) and (4.98) as

$$I_\odot(C^I) = I_\odot(\tilde{B}) = 1.333 > 0 = e,$$

hence, C^I is weak $\odot$-inconsistent.

4.8 Conclusion

This chapter deals with pairwise comparisons matrices with fuzzy elements and intuitionistic fuzzy elements. "Fuzzy" and/or "Intuitionistic fuzzy" entries of the pairwise comparisons matrix are applied whenever the decision-maker is not sure about the value of his/her evaluation of the relative importance of elements in question, the sense of both belonging and/or not belonging to a fuzzy set. In comparison with the PC matrices investigated in the literature, here we investigate pairwise comparisons matrices with elements from an abelian linearly ordered group (alo-group) over a real interval. By this, we generalize the concept of reciprocity and consistency of pairwise comparisons matrices with intuitionistic fuzzy numbers (IFPC matrices). We also define the concept of the priority vector which is an extension of the well-known concept in the crisp case and which is used for ranking the alternatives. Such an approach allows for extending the additive, multiplicative, and also fuzzy approaches known from the literature. Moreover, we also solve the problem of measuring the inconsistency of IFPC matrices by defining corresponding indexes: the weak consistency grade and inconsistency index. A number of numerical examples are presented to illustrate the concepts and derived properties.

The concept of the consistency of FPC matrices has already been studied in the author's former works as well as in the literature on this subject (see e.g., [17, 18, 29, 31, 36, 39, 44]). Here, we define a fundamental concept of consistency based on alpha-cuts, namely, two concepts: the "weak consistency" and the stronger concept of "consistency" of FPC matrices are presented. We investigate their properties as well as some consequences to the problem of ranking the alternatives. In particular, we derive necessary and sufficient conditions for FPC matrices to be weak consistent and consistent. Weak consistency as well as consistency are useful, for example, when constructing membership functions of fuzzy elements of FPC matrices. Moreover, we also solve the problem of measuring the inconsistency of FPC matrices by defining corresponding indexes. The first index, called the consistency grade, $g(A)$, is the maximal α of the α-cut, such that the corresponding FPC matrix is α-$\odot$-consistent. Moreover, the inconsistency index $I(A)$ of the FPC matrix A was defined for measuring the coherence of the FPC matrix. Consequently, an FPC matrix A is either consistent, in which case $0 \leq g(A) \leq 1$, and the inconsistency index $I(A)$ is equal to the identity element e of $\mathscr{G}$, or, the FPC matrix is inconsistent, in which case $g(A) = 0$, and $I(A) > e$. We extend this approach for PC matrices with intuitionistic fuzzy elements. Some numerical examples are presented to illustrate the concepts and properties.

We proposed a new categorization of inconsistent pairwise comparisons matrices with respect to the satisfaction/violation of selected PCM properties, such as the fundamental selection (FS) condition, preservation of order preference (POP) condition, preservation of order of intensity of preference (POIP) condition, and reliability priority (RP) condition. A new nonlinear optimization problem for finding the weights (i.e., priority vectors) satisfying the aforementioned conditions is proposed such that the distance function between the given PCM matrix and the ratio matrix composed of the weights is minimized. Moreover, we presented important examples of alo-groups where the above optimization problem for finding the desirable weights can be solved by standard optimization methods, e.g., interior point methods, or gradient-type methods.

We also unify several approaches known from the literature (see e.g., [8, 26, 29, 31, 36, 43, 44]). By doing this we solve the problem of measuring the inconsistency of an IFPC matrix C^I by defining corresponding indexes. The first index, called the consistency grade, $g(C^I)$ is the maximal α of the α-cut, such that the corresponding IFPC matrix is α-consistent. Moreover, the inconsistency index I of the IFPC matrix is defined for measuring the fuzziness of this matrix by the distance of the IFPC matrix to the closest crisp consistent matrix. Consequently, an IFPC matrix is both crisp and consistent, in which case g is equal to 1 and the consistency index I is equal to the identity element e or it is inconsistent, where $g < 1$ and I is greater than the identity element e. Several numerical examples were presented to illustrate the concepts and derived properties.

References

1. Atanassov KT (1986) Intutionistic fuzzy sets. Fuzzy Sets Syst 20:87–96
2. Bana e Costa AA, Vasnick JA (2008) A critical analysis of the eigenvalue method used to derive priorities in the AHP. Eur J Oper Res 187(3):1422–1428

3. Bilgic T, Turksen IB (2000) Measurement of membership functions: theoretical and empirical work. In: Dubois D, Prade H (eds) Fundamentals of fuzzy sets. Kluwer Academic Publishers, New York, pp 195–227

4. Blankmeyer E (1987) Approaches to consistency adjustments. J Optim Theory Appl 154:479–488

5. Bourbaki N (1998) Algebra II. Springer, Heidelberg

6. Boyd S, Vandenberghe L (2004) Convex optimization. Cambridge University Press, Cambridge

7. Bozbura FT, Beskese A (2007) Prioritization of organizational capital measurement indicators using fuzzy AHP. Int J Approx Reason 44(2):124–147

8. Buckley JJ (1985) Fuzzy hierarchical analysis. Fuzzy Sets Syst 17(1):233–247

9. Buckley JJ et al (2001) Fuzzy hierarchical analysis revisited. Eur J Oper Res 129:48–64

10. Cavallo B, Brunelli M (2018) A general unified framework for interval pairwise comparison matrices. Int J Approx Reason 93:178–198

11. Cavallo B, D'Apuzzo L (2009) A general unified framework for pairwise comparison matrices in multicriteria methods. Int J Intell Syst 24(4):377–398

12. Cavallo B, D'Apuzzo L (2012) Deriving weights from a pairwise comparison matrix over an alo-group. Soft Comput 16:353–366

13. Csutora R, Buckley JJ (2001) Fuzzy hierarchical analysis: the Lambda-Max method. Fuzzy Sets Syst 120:181–195

14. Gavalec M, Ramik J, Zimmermann K (2014) Decision making and optimization special matrices and their applications in economics and management. Springer International Publishing, Switzerland

15. Ishizaka A, Nguyen NH (2013) Calibrated fuzzy AHP for current bank account selection. Expert Syst Appl 40:3775–3783

16. Koczkodaj WW (1993) A new definition of consistency of pairwise comparisons. Math Comput Model 18:79–93

17. Kou G, Ergu D, Lin AS, Chen Y (2016) Pairwise comparison matrix in multiple criteria decision making. Technol Econ Dev Econ 22(5):738–765

18. Krejci J (2017) Fuzzy eigenvector method for obtaining normalized fuzzy weights from fuzzy pairwise comparison matrices. Fuzzy Sets Syst 315:26–43

19. Kulak O, Kahraman C (2005) Fuzzy multi-attribute selection among transportation companies using axiomatic design and analytic hierarchy process. Inf Sci 170:191–210

20. Kułakowski K (2015) A heuristic rating estimation algorithm for the pairwise comparisons method. Cent Eur J Oper Res 23(1):187–203

21. Laarhoven PJV, Pedrycz W (1983) A fuzzy extension of Saaty's priority theory. Fuzzy Sets Syst 11:199–227

22. Lee KL, Lin SC (2008) A fuzzy quantified SWOT procedure for environmental evaluation of an international distribution center. Inf Sci 178(2):531–549

23. Leung LC, Cao D (2000) On consistency and ranking of alternatives in fuzzy AHP. Eur J Oper Res 124:102–113

24. Li KW, Wang ZJ, Tong X (2016) Acceptability analysis and priority weight elicitation for interval multiplicative comparison matrices. Eur J Oper Res 250(2):628–638

25. Mahmoudzadeh M, Bafandeh AR (2013) A new method for consistency test in fuzzy AHP. J Intell Fuzzy Syst 25(2):457–461

26. Mikhailov L (2003) Deriving priorities from fuzzy pairwise comparison judgments. Fuzzy Sets Syst 134:365–385

27. Mikhailov L (2004) A fuzzy approach to deriving priorities from interval pairwise comparison judgements. Eur J Oper Res 159:687–704

28. Mikhailov L, Tsvetinov P (2004) Evaluation of services using a fuzzy analytic hierarchy process. Appl Soft Comput 5:23–33

29. Ohnishi S, Dubois D et al (2008) A fuzzy constraint based approach to the AHP. Uncertainty and intelligent information systems. World Scientific, Singapore, pp 217–228

30. Ramík J (2014) Isomorphisms between fuzzy pairwise comparison matrices. Fuzzy Optim Decis Mak 14:199–209

31. Ramík J (2015) Pairwise comparison matrix with fuzzy elements on alo-group. Inf Sci 297:236–253
32. Ramík J (2018) Strong reciprocity and strong consistency in pairwise comparison matrix with fuzzy elements. Fuzzy Optim Decis Mak 17:337–355
33. Ramík J, Korviny P (2010) Inconsistency of pairwise comparison matrix with fuzzy elements based on geometric mean. Fuzzy Sets Syst 161:1604–1613
34. Ramík J, Vlach M (2001) Generalized concavity in optimization and decision making. Kluwer Academic Publishers, Boston
35. Saaty TL (1991) Multicriteria decision making the analytical hierarchy process, vol I. RWS Publications, Pittsburgh
36. Salo AA (1996) On fuzzy ratio comparison in hierarchical decision models. Fuzzy Sets Syst 84:21–32
37. Vaidya OS, Kumar S (2006) Analytic hierarchy process: an overview of applications. Eur J Oper Res 169(1):1–29
38. Van Laarhoven PJM, Pedrycz W (1983) A fuzzy extension of Saaty's priority theory. Fuzzy Sets Syst 11(4):229–241
39. Wang ZJ (2015) Consistency analysis and priority derivation of triangular fuzzy preference relations based on modal value and geometric mean. Inf Sci 314:169–183
40. Wang ZJ, Li K (2015) A multi-step goal programming approach for group decision making with incomplete interval additive reciprocal comparison matrices. Eur J Oper Res 242(3):890–900
41. Whitaker R (2007) Criticisms of the analytic hierarchy process: why they often make no sense. Math Comput Model 46(7/8):948–961
42. Xu ZS (2012) Intuitionistic fuzzy aggregation and clustering. Studies in fuzziness and soft computing, vol 279. Springer, Heidelberg
43. Xu ZS (2014) Intuitionistic preference modeling and interactive decision making. Studies in fuzziness and soft computing, vol 280. Springer, Heidelberg
44. Xu ZS, Chen J (2008) Some models for deriving the priority weights from interval fuzzy preference relations. Eur J Oper Res 184:266–280
45. Xu ZS, Chen J (2008) An overview of distance and similarity measures of intuitionistic fuzzy sets. Int J Uncertain Fuzziness Knowl-Based Syst 16(4):529–555
46. Xu ZS, Hu H (2009) Entropy-based procedures for intuitionistic fuzzy multiple attribute decision making. J Syst Eng 20(5):1001–1011
47. Zhang H (2016) Group decision making based on multiplicative consistent reciprocal preference relations. Fuzzy Sets Syst 282:31–46

Chapter 5
Stochastic Approaches to Pairwise Comparisons Matrices in Decision-Making

5.1 Introduction

In this chapter we again deal with the pairwise comparisons matrices that we have already investigated in Chaps. 2, 3, and 4. However, here, we shall investigate PCMs with elements with a stochastic nature and particularly those that are random variables.

On the one hand, uncertain elements of the pairwise comparisons matrix could be obtained whenever the decision-maker is not sure about the preference degree of his/her evaluation of elements in question. Here, each element is given individually by the DM in the form of its membership function of a fuzzy set, which is subjectively evaluated and which is not a result of a repeatable event.

On the other hand, each element of the PCM is taken as data, observed values of random values from the set of real numbers with the given probability distribution of these values. The data are usually the outputs of a repetitive event in some stochastic situation. Then appropriate statistical methods are applied to the given data.

A *decision-making problem (DM problem)* is formulated again traditionally as follows, see Chap. 2: Let $\mathscr{C} = \{c_1, c_2, \cdots, c_n\}$ be a finite set of alternatives ($n > 2$). The aim is to rank the alternatives from the best to the worst (or vice versa), using the information given by a DM in the form of $n \times n$ PC matrix $\check{A} = \{\check{a}_{ij}\}$.

An ordinal *ranking* of alternatives is required to obtain the best alternative(s), however, it often occurs that the DM is not satisfied with the ordinal ranking among alternatives, and therefore a cardinal ranking, i.e., *rating* is required.

In Chap. 4, we have been dealing with the PCM $\tilde{A} = \{\tilde{a}_{ij}\}$, where each element is a bounded fuzzy interval of the alo-group $\mathscr{G}$. Here, in Chap. 5, by contrast, when applying statistical methods, a single group operation $\odot$ is not sufficient, and we need the other operation, e.g., not only multiplication ".", but also addition "+" is necessary. That is why we investigate here PCMs in the multiplicative system, i.e., a multiplicative alo-group within a ring structure of the set of real numbers $\mathbf{R}$ (see Example 4.2).

J. Ramik, *Pairwise Comparisons Method*, Lecture Notes in Economics and Mathematical Systems 690, https://doi.org/10.1007/978-3-030-39891-0_5

In the recent literature, we can find papers dealing with stochastic approaches to pairwise comparisons dated even in the 1950s and 1960s, a few decades earlier than the corresponding fuzzy approaches have emerged. The earliest work on PCMs using random quantities as data was published by Fechner [10] in 1856 and by Thurstone [26] in 1927. Most of the works on this topic have been published in 1950s ad 1960s (see e.g. [2, 5, 10, 19, 20, 23]). The recent works have emerged after year 2000, see e.g., [1, 6–9, 11, 13, 15, 16, 21, 22, 24, 25]. The seminal book on stochastic approaches in pairwise comparisons method by David [4], was originally published in 1963 with a second edition in 1988. This monograph deals with pairwise comparisons (named as "paired comparisons") in a broader sense, comprising also tournaments and other combinatorial methods as well as rich material about statistical testing of parameters of pairwise comparisons models. In this chapter, however, we deal with pairwise comparisons in a narrow sense, investigating only problems that are parallel to the problems of Chaps. 2, 3, and 4. The combinatorial methods, tournaments and statistical testing of parameters of the models are not investigated here.

5.2 Basic Models

Following the discussion in the monograph by David [4], there are a number of possible models which impose severe restrictions on the preference structure of probabilities. Let us consider n alternatives (i.e., variants, objects, treatments, stimuli, criteria, players, etc.) $c_1, c_2, \cdots, c_n$ to be compared in pairs by each of d judges, the k-th judge making r_k replications of all possible $\frac{n(n-1)}{2}$ comparisons. Let

$$x_{ijk\delta}, \ i, j \in \{1, \cdots, n\}, k \in \{1, \cdots, d\}, \delta \in \{1, \cdots, r_k\},$$

be an "indicator" random variable taking the values 0 or 1 according to whether the kth judge prefers c_i or c_j, when making the δth comparison of the two. Moreover, we assume that all comparisons are statistically independent, except for the condition

$$x_{ijk\delta} + x_{jik\delta} = 1.$$

Let us denote

$$\pi_{ijk\delta} = \Pr(x_{ijk\delta} = 1), \tag{5.1}$$

then the only restrictions are

$$0 \le \pi_{ijk\delta} \le 1, \text{ and } \pi_{jik\delta} = 1 - \pi_{ijk\delta}. \tag{5.2}$$

We consider the following special cases:

$$\pi_{ijk\delta} = \pi_{ij\delta}\text{—no replication effect,} \tag{5.3}$$

$$\pi_{ijk\delta} = \pi_{ij}\text{---no replication and judge effect.} \qquad (5.4)$$

If the replication effect is negligible, but differences between judges are not, model (5.4) is still very general as a description of individual preferences ($k = 1$). When applying model (5.4), the alternatives may be ranked according to the values of the average preference probabilities of c_i

$$\pi_{i.} = \frac{1}{n-1} \sum_{\substack{j=1 \\ j \neq i}}^{n} \pi_{ij}, \quad i \in \{1, \cdots, n\}. \qquad (5.5)$$

Here, $\pi_{i.} \geq \pi_{j.}$ implies that c_i "is better than" c_j.

Another kind of ranking is possible if the following *stochastic transitivity* holds for every triad of different alternatives c_i, c_j, c_k:

$$\pi_{ij} \geq \frac{1}{2}, \pi_{jk} \geq \frac{1}{2}, \pi_{ik} \geq \frac{1}{2}, \quad ij, k \in \{1, \cdots, n\}. \qquad (5.6)$$

Condition (5.6) leads to the total ordering of alternatives, compare with the definition of $\odot$-transitivity in Definition 3.10 and Proposition 3.12. Notice that stochastic transitivity (5.6) is equivalent to the multiplicative transitivity of PC matrix $\Pi = \{\pi_{ij}\}$ on the fuzzy multiplicative alo-group $\mathscr{R}_m$ (see Example 4.4).

5.3 Linear Models

In the traditional setting for pairwise comparisons, d decision-makers are invited to compare n alternatives $c_1, c_2, \cdots, c_n$ pairwise with respect to a single criterion and to state their preference for each pair (with no ties permitted). The results can be assembled in a matrix $D = \{d_{ij}\}$ where d_{ij} is the number of decision-makers who prefer alternative c_i to alternative c_j, for $i, j \in \{1, \cdots, n\}$ and thus $d_{ij} + d_{ji} = d$.

A full account of the analysis of such data is given in the monograph by David [4].

In what follows, strong conditions concerning the probability distributions of the random variables are assumed as follows. The perceived underlying merit of alternative c_i is taken to be a continuous random variable Y_i with mean μ_i. The probability that c_i is preferred to c_j can be introduced as

$$\pi_{ij} = P(c_i \succ c_j) = P(Y_i > Y_j), i, j \in \{1, \cdots, n\}, i \neq j. \qquad (5.7)$$

By defining

$$Z_i - Z_j = (Y_i - \mu_i) - (Y_j - \mu_j), \qquad (5.8)$$

and invoking some distributional arguments, it can be shown that

$$\pi_{ij} = P(Y_i > Y_j) = P(Z_i - Z_j > -(\mu_i - \mu_j)) = P(Z_i - Z_j < \mu_i - \mu_j) = H(\mu_i - \mu_j),$$
$$(5.9)$$

where $H(.)$ is the probability distribution function of $Z_i - Z_j$. Note that the merits
of the alternatives c_i, with $i \in \{1, \cdots, n\}$, are chosen on a *linear scale*. As the origin
of the linear scale is arbitrary, it is usual to impose an additional constraint on the μ_i
such that

$$\sum_{i=1}^{n} \mu_i = 0.$$

It also follows immediately from result (5.9) that $\pi_{ij} = \frac{1}{2}$ if and only if $\mu_i = \mu_j$,
that $\pi_{ij} > \frac{1}{2}$ if and only if $\mu_i > \mu_j$ and that $\pi_{ij} < \frac{1}{2}$ if and only if $\mu_i < \mu_j$ for
$i, j \in \{1, \cdots, n\}$ and $i \neq j$. Let

$$\delta_{ij} = \mu_i - \mu_j, i, j \in \{1, \cdots, n\}, \qquad (5.10)$$

so that the preference probability π_{ij} is given by

$$\pi_{ij} = H(\delta_{ij}), i, j \in \{1, \cdots, n\}, \qquad (5.11)$$

where H is the distribution function pertaining to the *linear model* chosen. Summing
up (5.10) over all j other than i, we obtain for $i \in \{1, \cdots, n\}$

$$\mu_i = \frac{1}{n} \sum_{j \neq i} \delta_{ij}. \qquad (5.12)$$

Formula (5.12) suggests the following method for finding estimates m_i of μ_i, i.e.,
the ranking of alternatives c_i, $i \in \{1, \cdots, n\}$, correspondingly.

For a balanced pairwise comparisons experiment with d replications, let

$$p_{ij} = \frac{d_{ij}}{d}$$

be the proportion of preferences for alternative c_i over c_j, i.e., $c_i \succ c_j$, with $i, j \in$
$\{1, \cdots, n\}$. As in (5.11), define f_{ij} for $i, j \in \{1, \cdots, n\}$ by

$$f_{ij} = H^{-1}(p_{ij}), \qquad (5.13)$$

then $f_{ji} = -f_{ij}$.

In general, it is impossible to satisfy all the relations $m_i - m_j = f_{ij}$ corresponding
to (5.10), since there are more equations than unknowns. However, these relations
can be, however, satisfied "on the average" in the sense of (5.12) by setting for
$i \in \{1, \cdots, n\}$

$$m_i = \frac{1}{n} \sum_{j \neq i} f_{ij}. \qquad (5.14)$$

The m_i obtained in this way are LSQ estimates of the μ_i, i.e., they minimize

$$S = \sum_{j \neq i} (f_{ij} - \mu_i + \mu_j)^2, \tag{5.15}$$

hence

$$\frac{\partial S}{\partial \mu_i} = -2 \sum_{j \neq i} (f_{ij} - \mu_i + \mu_j) + 2 \sum_{j \neq i} (f_{ji} - \mu_j + \mu_i) =$$

$$= -4 \sum_{j \neq i} (f_{ij} - \delta_{ij}) = 0, \tag{5.16}$$

which by (5.12) gives (5.14) as the solution of (5.16).

Clearly, the other estimates of the μ_i are possible, but the simple estimates (5.14) have the advantage that the method of computations remains the same whatever the form of H may be. All we need is an appropriate algorithm (or computer program) for the distribution function H or, even better, for H^{-1}, so that given p_{ij}, we obtain f_{ij} by (5.13).

5.3.1 Thurstone–Mosteller Model

The preference probabilities π_{ij}, for $i, j \in \{1, ..., n\}$ and $i \neq j$ given in Eq.(5.9) are specified by the choice of distribution function H and two such choices are of particular interest: the Thurstone–Mosteller model [23], and Bradley–Terry model, [2], which will be investigated in what follows.

The variables Y_i are taken to be normally distributed and then it follows that

$$\pi_{ij} = P(Y_i > Y_j) = \Phi(\frac{\mu_i - \mu_j}{\sigma}), \tag{5.17}$$

where $\Phi(.)$ is the cumulative distribution function of the standard normal distribution and σ^2 is the variance of the difference $Y_i - Y_j$. Moreover, we obtain

$$\pi_{ij} = H(\mu_i - \mu_j) = \int_{-\frac{\mu_i - \mu_j}{\sigma}}^{\infty} z(x)dx, \tag{5.18}$$

where

$$z(x) = \frac{1}{\sqrt{2\pi}} e^{-\frac{1}{2}x^2}. \tag{5.19}$$

Example 5.1 Consider $d = 10$ decision-makers, who compare four alternatives: c_1, c_2, c_3, c_4 by pairwise comparisons with respect to a single criterion. The results can be assembled in a matrix $P = \{p_{ij}\}$ where p_{ij} is the relative number of decision-makers who prefer alternative c_i to alternative c_j, for $i, j \in \{1, 2, 3, 4\}$ and thus $p_{ij} + p_{ji} = 1$.

$$P = \{p_{ij}\} = \begin{pmatrix} 0 & 0.7 & 0.4 & 0.2 \\ 0.3 & 0 & 0.5 & 0.3 \\ 0.6 & 0.5 & 0 & 0.1 \\ 0.8 & 0.7 & 0.9 & 0 \end{pmatrix}. \tag{5.20}$$

Then, by (5.13) we obtain $F = \{f_{ij}\}$ by applying the inverse cumulative distribution function of the standard normal distribution (5.18), (5.19):

$$F = \{f_{ij}\} = \begin{pmatrix} 0 & 0.524 & -0.253 & -0.842 \\ -0.524 & 0 & 0 & -0.524 \\ 0.253 & 0 & 0 & -1.282 \\ 0.842 & 0.524 & 1.282 & 0 \end{pmatrix}. \tag{5.21}$$

By (5.14) we obtain m_i, and the estimates of μ_i, $i \in \{1, 2, 3, 4\}$, as follows:

$$m = (m_1, m_2, m_3, m_4) = (-0.143, -0.262, -0.257, 0.662).$$

Hence, we obtain the corresponding ranking of alternatives:

$$c_4 \succ c_1 \succ c_3 \succ c_2.$$

5.3.2 Bradley–Terry Model

In the more widely used Bradley–Terry model [2], the variables Y_i are taken to follow independent distribution functions and the differences $Y_i - Y_j$, therefore, follow logistic distributions [5]:

$$\pi_{ij} = H(\mu_i - \mu_j) = \frac{1}{1 + e^{-\frac{\mu_i - \mu_j}{s}}}, \quad i, j \in \{1, \cdots, n\}. \tag{5.22}$$

It can then be shown that

$$\pi_{ij} = P(Y_i > Y_j) = \frac{\pi_i}{\pi_i + \pi_j}, \tag{5.23}$$

where the parameters $\pi_i = \frac{1}{n} \sum_{j \neq i} \pi_{ij}$ have an immediate and natural interpretation as weights or probabilities associated with the objects c_i, with $i \in \{1, \cdots, n\}$, respectively.

Suppose now that n alternatives are again to be compared according to a particular criterion but that a single DM expresses his/her relative preferences on a ratio scale, now, evaluated repeatedly now, or eventually, by a number of judges (i.e., DMs). Saaty in [21] suggested choosing relative preferences on an ordinal scale of integers from 1 to 9, together with their reciprocals, for details see Chap. 2. However, other scales can also be used. For example, Becker et al. [1], used a five-point Likert scale. More formally, let a_{ij} denote the relative preference of alternative c_i when compared with alternative c_j, $j \in \{1, \cdots, n\}$. The relative preferences can then be assembled in a positive reciprocal matrix, termed a *pairwise comparisons matrix (PCM)*, of the form $A = \{a_{ij}\}$ where

$$a_{ij} > 0, a_{ji} = \frac{1}{a_{ij}}, \text{ and } a_{ii} = 1 \text{ for } i, j \in \{1, \cdots, n\}.$$

The entries of matrix A can be regarded as the ratios of weights associated with the objects in the pairwise comparisons. Once the pairwise judgments have been elicited, the crucial question is how to determine the weights associated with the objects. Here, we deal with three *distributional approaches*.

5.3.3 Logarithmic Least Squares and the Normal Distribution

Crawford and Williams in [3] introduced probably the first statistical approach to the analysis of pairwise comparisons matrices, termed the *logarithmic least squares method (LLSM)*, in 1985. A more detailed description on the LLSM was presented in Chap. 2, Sect. 2.5.4. See also Kabera and Haines [16] and Laininen and Hamalainen [22]. The LLSM can, however, be derived using the same arguments as those invoked in deriving the Thurstone–Mosteller model for pairwise comparisons. The relative preferences a_{ij}, an observed value of the random value $\hat{a}_{ij}$, i.e., data, can be expressed as

$$a_{ij} = \frac{w_i}{w_j} e_{ij}^*, \tag{5.24}$$

where w_i and w_j are the unknown weights associated with the alternative c_i and alternative c_j, respectively. Moreover, $w_i > 0$ for $i \in \{1, ..., n\}$ and $\sum_{i=1}^{n} w_i = 1$, and e_{ij}^* is a positive error which captures the inconsistency in the judgments.

By invoking a logarithmic transformation, model (5.24) can be expressed as the linear model without intercept as follows:

$$y_{ij} = \beta_i - \beta_j + e_{ij}, 1 \le i < j < n, \tag{5.25}$$

where

$$y_{ij} = \ln a_{ij}, \beta_i = \ln w_i, \beta_j = \ln w_j, e_{ij} = \ln e_{ij}^*. \tag{5.26}$$

Assuming that the differences y_{ij} are observed values of random variables $Y_{ij} = Y_i - Y_j$ taken to be normally distributed (as in the Thurstone–Mosteller model), that is $Y_{ij} \sim N(\mu_i - \mu_j, \sigma^2)$, simple arguments yield the following results. Pairwise differences of the parameters, $\mu_i - \mu_j$ can be estimated and it is straightforward to show that the least squares or, equivalently, the maximum likelihood estimates (MLEs) are given by

$$\hat{\beta}_i - \hat{\beta}_j = \frac{1}{n}(2y_{ij} + \sum_{\substack{k=1 \\ k \neq i,j}}^{n} y_{ik} - \sum_{\substack{k=1 \\ k \neq i,j}}^{n} y_{jk}), \quad 1 \leq i < j \leq n. \qquad (5.27)$$

Estimates of the weights associated with the alternatives can be expressed as

$$\hat{w}_i = \frac{\exp(\hat{\beta}_i)}{\sum_{j=1}^{n} \exp(\hat{\beta}_j)} = \frac{1}{\sum_{j=1}^{n} \exp[-(\hat{\beta}_i - \hat{\beta}_j)]}, \quad i \in \{1, \cdots, n\}. \qquad (5.28)$$

Since these weights depend only on the unique estimates of $\beta_i - \beta_j$, they are unique. Approximate variances and covariances of the estimates $\hat{w}_i$ for $i \in \{1, \cdots, n\}$, $Var(\hat{w}_i)$ and $Cov(\hat{w}_i, \hat{w}_j)$, may be obtained, for example, in [16].

Example 5.2 Consider data from Example 5.1, i.e., $d = 10$ decision-makers, who compare four alternatives: c_1, c_2, c_3, c_4 by pairwise comparisons. The results can be assembled in a PC matrix $A = \{a_{ij}\}$ where a_{ij} are the results of relative preferences of decision-makers who prefer alternative c_i to alternative c_j, for $i, j \in \{1, 2, 3, 4\}$ on the nine-point scale by Saaty. Each judge $\gamma \in \{1, \cdots, 10\}$ evaluates his/her preference between alternative c_i and c_j by the value $a_{ij\gamma}$. Then the element a_{ij} of PCM $A = \{a_{ij}\}$ is calculated as a product of all $a_{ij\gamma}$, specifically,

$$a_{ij} = \prod_{\gamma=1}^{10} a_{ij\gamma}.$$

The resulting PCM $A = \{a_{ij}\}$ is reciprocal, which means that $a_{ji} = \frac{1}{a_{ij}}$, as all the evaluations of the judges are also reciprocal. For more details, see Chap. 2. The other approaches will be dealt with in Sect. 5.4.1.

Let $A = \{a_{ij}\}$ be a given PCM as follows:

$$A = \{a_{ij}\} = \begin{pmatrix} 1 & 2 & 3 & 5 \\ \frac{1}{2} & 1 & 2 & 3 \\ \frac{1}{3} & \frac{1}{2} & 1 & 2 \\ \frac{1}{5} & \frac{1}{3} & \frac{1}{2} & 1 \end{pmatrix}. \qquad (5.29)$$

By (5.28) and (5.25), (5.26), we obtain w_i, the estimates of w_i, by the LLSQ method as follows:

$$w = (w_1, w_2, w_3, w_4) = (0.483, 0.272, 0.157, 0.088).$$

Hence, we obtain the corresponding ranking of alternatives:

$$c_1 \succ c_2 \succ c_3 \succ c_4.$$

5.4 Direct Approaches

Now, we deal with the question as to whether the previous pairwise comparisons models can be used more directly in modeling PC matrices of the form $A = \{a_{ij}\}$ than the models discussed above.

Genest and M'Lan in [12] suggested that a_{ij}, the relative preference for alternative c_i with respect to alternative c_j, recorded on a ratio scale such as the Saaty's scale $\{\frac{1}{9}, \frac{1}{8}, \cdots, 1, 2, \cdots, 8, 9\}$, can be interpreted as reflecting the fact that the alternatives have been compared n_{ij} times where

$$n_{ij} = \max\{a_{ij}, a_{ji}\} + 1, 1 \le i < j \le n, \tag{5.30}$$

and thus that c_i is preferred to c_j totally x_{ij} times where x_{ij} is determined by

$$a_{ij} = \frac{x_{ij}}{n_{ij} - x_{ij}}.$$

Hence,

$$x_{ij} = \frac{n_{ij} a_{ij}}{1 + a_{ij}},$$

or, more succinctly,

$$x_{ij} = \max\{a_{ij}, 1\}, 1 \le i < j \le n.$$

With this interpretation, the pairwise comparison of alternatives c_i and c_j can be regarded as a binomial distribution with the number of trials n_{ij} taken to be independent, the number of successes given by x_{ij} and the probability of success π_{ij} taken to follow the Bradley–Terry model for pairwise comparisons.

Here, we present a popular approach, namely, when the entry a_{ij} in a PC matrix $A = \{a_{ij}\}$ is interpreted as an odds ratio for the preference probability, and thus as

$$a_{ij} = \frac{p_{ij}}{1 - p_{ij}} \text{ for } 1 \le i < j \le n, \tag{5.31}$$

or, equivalently,

$$p_{ij} = \frac{a_{ij}}{a_{ij} + 1} \text{ for } 1 \le i < j \le n, \tag{5.32}$$

where p_{ij} is the observed probability that alternative c_i is preferred to alternative c_j, for $1 \le i < j \le n$. Note that the relative preference a_{ij} is taken to be on a ratio scale, for example, (but not necessarily) the Saaty's scale. Estimates of the parameters of models describing the preference probabilities π_{ij} can be obtained as those values for which the observed probabilities p_{ij} are as close as possible, in some sense, to the true values $\pi_{ij}, 1 \leq i < j \leq n$.

Two measures of "closeness", one based on least squares and the other on the Kullback–Leibler distance, are now considered and the ideas reinforced by invoking the Bradley–Terry model (5.23). Consider the LSQ approach of David, see [4], which involves minimizing the sum of squares, (5.15) as

$$S = \sum_{\substack{j \neq i \\ j=1}}^{n} [H^{-1}(p_{ij}) - H^{-1}(\pi_{ij})]^2 = \sum_{\substack{j \neq i \\ j=1}}^{n} [H^{-1}(p_{ij}) - (\mu_i - \mu_j)]^2, \qquad (5.33)$$

where the observed probabilities p_{ij} are given by Eq. (5.9) and $H(.)$ is given by (5.17). For the Bradley–Terry model (5.23), $\mu_i = \ln(\pi_i)$ and

$$H^{-1}(p_{ij}) = \ln \frac{p_{ij}}{1 - p_{ij}} = \ln a_{ij}, \, 1 \leq i < j \leq n, \qquad (5.34)$$

so that estimates of the parameters are obtained by minimizing the expression

$$\sum_{\substack{j \neq i \\ j=1}}^{n} (\ln a_{ij} - \ln \frac{\pi_i}{\pi_j})^2. \qquad (5.35)$$

Note that this approach applied to the Bradley–Terry model (5.23) gives the same results as the LLSM applied to the Thurstone–Mosteller model (5.27) discussed earlier in Sect. 5.2, although under different assumptions and also different interpretation, (see also [16, 22]).

Example 5.3 Consider data from Example 5.1, i.e., $d = 10$ decision-makers, who compare four alternatives: c_1, c_2, c_3, c_4 by pairwise comparisons. The results can be assembled in a matrix $P = \{p_{ij}\}$ where p_{ij} is the relative number of decision-makers who prefer alternative c_i to alternative c_j, for $i, j \in \{1, 2, 3, 4\}$ and thus $p_{ij} + p_{ji} = 1$.

$$P = \{p_{ij}\} = \begin{pmatrix} 0 & 0.7 & 0.4 & 0.2 \\ 0.3 & 0 & 0.5 & 0.3 \\ 0.6 & 0.5 & 0 & 0.1 \\ 0.8 & 0.7 & 0.9 & 0 \end{pmatrix}. \qquad (5.36)$$

Now, we construct a PC matrix $A = \{a_{ij}\}$, where the entry a_{ij} is interpreted as an odds ratio for the preference probability, and thus as

$$a_{ij} = \frac{p_{ij}}{1 - p_{ij}} \text{ for } 1 \leq i < j \leq 4, \qquad (5.37)$$

Then $A = \{a_{ij}\}$ is as follows:

$$A = \{a_{ij}\} = \begin{pmatrix} 1 & \frac{7}{3} & \frac{2}{3} & \frac{1}{4} \\ \frac{3}{7} & 1 & 1 & \frac{3}{7} \\ \frac{3}{2} & 1 & 1 & \frac{1}{9} \\ 4 & \frac{7}{3} & 9 & 1 \end{pmatrix}. \tag{5.38}$$

Applying the B–T model (5.23) and LSM, we obtain the following estimates of probabilities:

$$\pi = (\pi_1, \pi_2, \pi_3, \pi_4) = (0.155, 0.128, 0.125, 0.592).$$

Hence, we obtain the corresponding ranking of alternatives:

$$c_4 \succ c_1 \succ c_3 \succ c_2,$$

which is the same ranking as in Example 5.1.

5.4.1 The Kullback–Leibler Distance

The *Kullback–Leibler distance (K–L distance)*, D, between the probabilities p_{ij} and π_{ij} for $1 \le i < j \le n$, with respect to the distribution specified by the p_{ij} is given by

$$D = \sum_{\substack{j<i \\ i,j=1}}^{n} [p_{ij} \ln \frac{p_{ij}}{\pi_{ij}} + (1 - p_{ij}) \ln \frac{1 - p_{ij}}{1 - \pi_{ij}}], \tag{5.39}$$

see [19, 20].

Here, p_{ij} are observed (i.e., given) probabilities, and therefore, minimizing D is equivalent to maximizing D^* as follows:

$$D^* = \sum_{\substack{j<i \\ i,j=1}}^{n} [p_{ij} \ln \frac{\pi_{ij}}{1 - \pi_{ij}} + \ln(1 - \pi_{ij})]. \tag{5.40}$$

Moreover, maximizing expression (5.40) is equivalent to maximizing the log-likelihood function for a binomial distribution with numbers of trials all equal and with probabilities of success π_{ij}. In particular, note that the model proposed by Genest and M'Lan [12] gives the same parameter estimates as those obtained by maximizing (5.40) provided that all pairs of alternatives are assumed to be compared the same number of times. The present approach, therefore, gives some support to their methodology.

Consider now the case where the probability π_{ij} that alternative c_i is preferred to alternative c_j follows the Bradley–Terry model (5.23), that is,

$$\pi_{ij} = \frac{\pi_i}{\pi_i + \pi_j}, \quad i, j \in \{1, \cdots, n\}, \tag{5.41}$$

and

$$D^* = \sum_{\substack{j<i \\ i,j=1}}^{n} [p_{ij} \ln \frac{\pi_i}{\pi_j} + \ln \frac{\pi_i}{\pi_i + \pi_j}]. \tag{5.42}$$

Estimates of the weights which minimize the K–L distance (5.39) can be readily obtained by developing an iterative procedure that preserves the constraints that $\pi_i > 0$ and $\sum_{i=1}^{n} \pi_i = 1$ and that is similar to the one developed for the method of pairwise comparisons [4]. Specifically, using (5.42) and solving for $\frac{\partial D^*}{\partial \pi_i} = 0$ gives

$$\frac{1}{\hat{\pi}_i} \sum_{j=i+1}^{n} p_{ij} = \sum_{j=i+1}^{n} (\hat{\pi}_i + \hat{\pi}_j)^{-1}, \quad i \in \{1, \cdots, n\}. \tag{5.43}$$

where $\hat{\pi}_i$ is an estimate of π_i. This system of equations then forms the basis for an iterative scheme for finding the estimates $\hat{\pi}_i, i \in \{1, \cdots, n\}$. The calculations require the following algorithm, [16]:

Step 1. Choose starting values for the $\hat{\pi}_i$, such as

$$\hat{\pi}_1^{(0)}, \hat{\pi}_2^{(0)}, ..., \hat{\pi}_n^{(0)} = \frac{1}{n}. \tag{5.44}$$

Step 2. Take the kth iteration, $k = 1, 2, ...,$ as

$$\hat{\pi}_i^{(k)} = \frac{\sum_{j=i+1}^{n} p_{ij}}{\sum_{j=i+1}^{n} (\hat{\pi}_i^{(k-1)} + \hat{\pi}_j^{(k-1)})^{-1}}, \quad i \in \{1, \cdots, n\}. \tag{5.45}$$

Step 3. Continue the process until $\hat{\pi}_i^{(k+1)}$ is sufficiently close to $\hat{\pi}_i^{(k)}$, say at iteration N, and then the final estimates are taken to be

$$\hat{\pi}_i = \frac{\hat{\pi}_i^{(N)}}{\sum_{j=1}^{n} \hat{\pi}_j^{(N)}}, \quad i \in \{1, \cdots, n\}. \tag{5.46}$$

Alternatively the weights can be estimated by invoking a constrained optimization routine. The K–L distance is not a likelihood function. Thus standard errors for the parameter estimates cannot be obtained from the asymptotic results of likelihood theory. However, the jackknife technique can be used to estimate the weights and to provide attendant standard errors. In particular, consider estimating weights by omitting the comparison of alternative c_i and c_j as the vector $\hat{\pi}_{(i,j)}$, where the subscripts indicate the removed pairs for $i, j \in \{1, \cdots, n\}$. There are $\frac{n(n-1)}{2}$ such estimates and these form the basis for finding an overall estimate $\hat{\pi}_i$ and the associated standard errors.

Example 5.4 Consider data from Example 5.3, i.e., $d = 10$ decision-makers, who compare four alternatives: c_1, c_2, c_3, c_4 by pairwise comparisons. The results can be assembled in a matrix $P = \{p_{ij}\}$ where p_{ij} is the relative number of decision-makers who prefer alternative c_i to alternative c_j, for $i, j \in \{1, 2, 3, 4\}$:

$$P = \{p_{ij}\} = \begin{pmatrix} 0 & 0.7 & 0.4 & 0.2 \\ 0.3 & 0 & 0.5 & 0.3 \\ 0.6 & 0.5 & 0 & 0.1 \\ 0.8 & 0.7 & 0.9 & 0 \end{pmatrix}. \tag{5.47}$$

Now, we construct a PC matrix $A = \{a_{ij}\}$ as follows:

$$a_{ij} = \frac{p_{ij}}{1 - p_{ij}} \text{ for } 1 \le i < j \le 4, \tag{5.48}$$

Then $A = \{a_{ij}\}$ is as follows:

$$A = \{a_{ij}\} = \begin{pmatrix} 1 & \frac{7}{3} & \frac{2}{3} & \frac{1}{4} \\ \frac{3}{7} & 1 & 1 & \frac{3}{7} \\ \frac{3}{2} & 1 & 1 & \frac{1}{9} \\ 4 & \frac{7}{3} & 9 & 1 \end{pmatrix}. \tag{5.49}$$

Applying the L–K distance maximizing (5.42) and applying algorithm (5.44)–(5.46), we obtain the following estimates of probabilities:

$$\pi = (\pi_1, \pi_2, \pi_3, \pi_4) = (0.144, 0.062, 0.216, 0.577).$$

Hence, we obtain the corresponding ranking of alternatives:

$$c_4 \succ c_3 \succ c_1 \succ c_2,$$

which is a different ranking compared to Example 5.3.

5.5 Conclusion

This chapter deals with a number of stochastic approaches to pairwise comparisons matrix. In the context of the previous chapters, we deal again with pairwise comparisons of a finite set of alternatives. Fuzzy elements of a PCMs, investigated in the previous chapter, are usually applied whenever the decision-maker is not sure about the value of his/her evaluations of the relative importance of elements in question. The result of pairwise comparisons is mostly subjective and no repetitive event is assumed. Stochastic elements, investigated in this chapter, are usually applied whenever the element is random, which means that its values may vary in individual observations.

Moreover, the problems investigated here are closely associated with that of Chap. 2. On the one hand, the elements of each PCM can be taken as observed values of random matrix entries and as such any of the ranking elicitation methods discussed in Chap. 2 can be applied. On the other hand, the elements of each PCM can be results of statistical methods applied to the observed data. In both cases, the approach to deriving the ranking of the alternatives may be combined with the special method discussed in this chapter.

We first reviewed the key statistical approaches to extracting the weights of objects from a PC matrix, and then related the embedded models to the traditional linear stochastic models used in the method of pairwise comparisons. We gave a brief account of the method of pairwise comparisons and a formal specification of a PC matrix. Statistical approaches to the analysis of judgment matrices were then introduced within the context of pairwise comparisons, with those which are distribution-based, and those based more directly on the method of pairwise comparisons. Combinatorial methods, tournaments, and the statistical testing of parameters of the models are not investigated here. Illustrative examples were presented and discussed and conclusions and prospects for further research were given.

References

1. Becker PJ, Wolvaardt JS et al (2009) A composite score for a measuring instrument utilising re-scaled Likert values and item weights from matrices of pairwise ratios. J Interdiscip Health Sci 14(1):29–32
2. Bradley RA, Terry ME (1952) Rank analysis of incomplete block designs. I. The method of paired comparisons. Biometrika 39:324–345
3. Crawford G, Williams C (1985) A note on the analysis of subjective judgment matrices. J Math Psychol 29(4):387–405
4. David HA (1988) The method of paired comparisons, 2nd edn. Charles Griffin, London
5. Davidson RR (1969) On a relationship between two representations of a model for paired comparisons. Biometrics 25(3):597–599
6. Durbach I (2006) A simulation-based test of stochastic multicriteria acceptability analysis using achievement functions. Eur J Oper Res 170:923–934
7. Durbach I (2009) On the estimation of a satisficing model of choice using stochastic multicriteria acceptability analysis. Omega 37:497–509
8. Durbach I (2009) The use of the SMAA acceptability index in descriptive decision analysis. Eur J Oper Res 196:1229–1237
9. Fan Z, Liu Y, Feng B (2010) A method for stochastic multiple criteria decision making based on comparisons of alternatives with random evaluations. Eur J Oper Res 207:906–915
10. Fechner GT (1965) Elements of psychophysics. Holt, Reinhard and Winston, New York
11. Fouss F, Achbany Y, Saerens M (2010) A probabilistic reputation model based on transaction ratings. Inf Sci 180:2095–2123

12. Genest C, M'Lan CE (1999) Deriving priorities from the Bradley-Terry model. Math Comput Model 29(4):87–102
13. Graves SB, Ringuest JL (2009) Probabilistic dominance criteria for comparing uncertain alternatives: a tutorial. Omega 37:346–357
14. Ishizaka A, Labib A (2011) Review of the main developments in the analytic hierarchy process. Expert Syst Appl 38(11):14336–14345
15. Jones DF, Mardle SJ (2004) A distance-metric methodology for the derivation of weights from a pairwise comparison matrix. J Oper Res Soc 55(8):869–875
16. Kabera MG, Haines LM (2013) A note on the statistical analysis of point judgment matrices. ORiON 29(1):75–86
17. Kaya I, Kahraman C (2010) Development of fuzzy process accuracy index for decision making problems. Inf Sci 180:861–872
18. Keeney R, Raiffa H (1976) Decisions with multiple objectives: preferences and value tradeoffs. Wiley, New York
19. Kullback S, Leibler RA (1951) On information and sufficiency. Ann Math Stat 22(1):79–86
20. Kullback S (1959) Information theory and statistics. Wiley, New York (Republished by Dover Publications in 1968; 1978)
21. Lahdelma R, Salminen P, Makkonen S (2006) Multivariate Gaussian criteria in SMAA. Eur J Oper Res 170:957–970
22. Laininen P, Hamalainen RP (2003) Analyzing AHP-matrices by regression. Eur J Oper Res 148(3):514–524
23. Mosteller F (1977) Remarks on the method of paired comparisons: I. The least-squares solution assuming equal standard deviations and equal correlations. Psychometrika 16(1):3–9
24. Nowak M (2006) INSDECMan interactive procedure for stochastic multicriteria decision problems. Eur J Oper Res 175:1413–1430
25. Tervonen T, Lahdelma R (2007) Implementing stochastic multicriteria acceptability analysis. Eur J Oper Res 178:500–513
26. Thurstone LL (1927) A law of comparative judgment. Psychol Rev 344:273–289

Part II
Pairwise Comparisons Method—Applications in Decision Making

Chapter 6
Applications in Decision-Making: Analytic Hierarchy Process—AHP Revisited

6.1 Introduction

A fundamental problem of decision theory is how to derive weights for a set of criteria (activities, alternatives, objects, etc.) according to their importance. Importance is usually judged according to *several criteria* [3, 4]. Each criterion may be shared by some or all of the activities. The criteria may, for example, be objectives in which the activities have been devised to fulfill. This is a process of multiple criteria decision-making, which we study here through a theory of measurement in a hierarchical structure.

Currently, pairwise comparisons constitute the core of the analytic hierarchy process (AHP) and the analytic network process (ANP), popular theoretical frameworks for the multiple criteria decision-making proposed by Thomas L. Saaty, see [6, 7, 17, 18].

The analytic hierarchy process (AHP) is a theory of relative measurement on absolute scales of both tangible and intangible criteria based on the paired comparison judgment of knowledgeable experts. Comparisons can serve as a tool of measurement and enable a valid scale of priorities to be derived from these measurements. How to measure intangibles is the main concern of the mathematics of the AHP. But it must work with tangibles as well to give back measurements that can be used for the tangible factors in a decision with accuracy on tangibles. The AHP has been validated with numerous examples in applications (see the review paper [29]). The analytic hierarchy process (AHP) is a structured optimization technique for organizing and analyzing complex decisions, based on mathematics and psychology. It was developed by Thomas L. Saaty in the 1970s and has been extensively studied and refined since then ([17–27]). Currently, it represents the most accurate approach for quantifying the weights of criteria. Individual expert's experiences are utilized to estimate the relative magnitudes of factors through pairwise comparisons in the form of a pairwise comparisons matrix (PCM).

AHP is also understood as an optimization methodology for finding the "optimal" solution among a finite number of alternatives subject to a finite number of criteria of

J. Ramik, *Pairwise Comparisons Method*, Lecture Notes in Economics and Mathematical Systems 690, https://doi.org/10.1007/978-3-030-39891-0_6

quantitative as well as qualitative nature. The individual criteria may be characterized by the corresponding PCMs expressing the individual goals that should be satisfied. Then the individual goals are structured into the hierarchy covered by the global (main) goal at the top of the entire hierarchy.

The main advantage of using PCMs is when an explicit formulation of the criterion is not known. Then it allows the experts and decision-makers to compare two alternatives at a time, thus reducing the complexity of a decision-making problem, especially when the set of compared elements (e.g., alternatives) $\mathscr{C} = \{c_1, c_2, \cdots, c_n\}$ is large enough, and it represents a tool for deriving a priority vector (or weighting vector) that establishes a final ranking of the alternatives in $\mathscr{C}$ (see e.g., [1]).

Rather than prescribing an "optimal" decision, the AHP helps DMs find an alternative that best suits their global goal and their understanding of the problem. It provides a comprehensive and rational framework for structuring a decision problem, for representing and quantifying its elements, for relating those elements to local and global goals, and for evaluating possible solutions.

First, the users of the AHP decompose their decision problem into a hierarchy of more easily comprehended subproblems, each of which can be analyzed independently. Each element of the hierarchy is associated with a PCM corresponding to a simpler subproblem of generating the local priority vector. Here, one has to solve the subproblem that has been already investigated in the previous Chaps. 2–5.

Once the hierarchy is built, the decision-makers systematically evaluate its various elements by comparing them to each other, two at a time, with respect to their impact on an element above them in the hierarchy. In making the comparisons, the decision-makers can use concrete data about the elements, but they use their judgments about the elements' relative importance.

The AHP converts these evaluations to numerical values that can be processed and compared over the entire range of the problem. A numerical weight or priority is derived for each element of the hierarchy, allowing diverse and often incommensurable elements to be compared to one another in a rational and consistent way. This capability distinguishes the AHP from some other decision-making methods.

In the final step of the AHP process, numerical priorities are calculated for each of the decision alternatives. These numbers represent the alternatives' relative ability to achieve the main decision goal, so they allow a straightforward consideration of the various courses of action.

The AHP has unique advantages when important elements of the decisions are difficult to quantify or compare, or where communication among team members is impeded by their different specializations, terminologies, or perspectives.

6.2 Applications of AHP

Decision situations to which the AHP can be applied include [5]:

(i)   Optimal choice—The selection of one alternative from a given set of alternatives, usually where there are multiple decision criteria involved.

(ii)   Ranking—Setting a set of alternatives in order from most to least desirable.

(iii) Prioritization—Determining the relative merit of members of a set of alternatives, as opposed to selecting a single one or merely ranking them.

(iv) Resource allocation—Apportioning resources among a set of alternatives.

(v) Benchmarking—Comparing the processes in one's own organization with those of other best-of-breed organizations.

(vi) Quality management—Dealing with the multidimensional aspects of quality and quality improvement.

(vii) Conflict resolution—Settling disputes between parties with apparently incompatible goals or positions.

The applications of AHP to complex decision situations have numbered in the thousands (see [29]), and have produced extensive results in problems involving planning, resource allocation, priority setting, and selection among alternatives, see [2]. Other areas have included forecasting, total quality management, business process reengineering, quality function deployment, and the balanced scorecard. Many AHP applications are never reported to the world at large, because they take place at high levels of large organizations where security and privacy considerations prohibit their disclosure, see [6]. But some uses of AHP are discussed in the literature. These have included the following [29]:

- Selecting a type of nuclear reactors (Politecnico di Milano).
- Deciding how best to reduce the impact of global climate change (Fondazione Eni Enrico Mattei).
- Quantifying the overall quality of software systems (Microsoft Corporation).
- Selecting university faculty (Bloomsburg University of Pennsylvania).
- Deciding where to locate offshore manufacturing plants (University of Cambridge).
- Assessing risk in operating cross-country petroleum pipelines (American Society of Civil Engineers).
- Deciding how best to manage U.S. watersheds (U.S. Department of Agriculture).
- More effectively defining and evaluating SAP Implementation Approaches (SAP Experts).
- An accelerated bridge construction decision-making tool to assist in determining the viability of accelerated bridge construction (ABC) over traditional construction methods and in selecting appropriate construction and contracting strategies on a case-by-case basis.

6.3 Establishing Priorities

Priorities are values (numbers) associated with the nodes of an AHP hierarchy. They represent the relative weights of the nodes in any hierarchy element. Like probabilities, priorities are absolute numbers between zero and one, without units or dimensions. Depending on the problem at hand, "weight" can refer to importance, or preference, or likelihood, or whatever factor is being considered by the decision-makers.

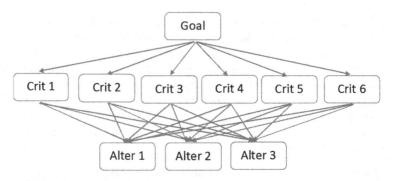

Fig. 6.1 Hierarchy with 4 criteria and 3 alternatives

Priorities are distributed over a hierarchy according to its architecture, and their values depend on the information entered by the users of the process. We distinguish priorities of the goal, the criteria, and the alternatives are closely related, but need to be considered separately. By definition, the priority of the goal is 1.000. That the priorities of the alternatives always add up to 1.000. Things can become complicated with multiple levels of criteria, but if there is only one level, their priorities also add up to 1.000. All this is illustrated by the priorities in Fig. 6.1. Here, we have a simple AHP hierarchy with associated default priorities.

Observe that the priorities on each level of the example—the goal, the criteria, and the alternatives all add up to 1.000.

The priorities shown in Fig. 6.1 are those that exist before any information has been entered about the weights of the criteria or alternatives, so the priorities within each level are all equal. They are called the hierarchy's default priorities. If a fifth criterion were added to this hierarchy, the default priority for each criterion would be now 0.200. If there were only two alternatives, each would have a default priority of 0.500.

Two additional concepts apply when a hierarchy has more than one level of criteria: *local priorities* and *global priorities*. Consider the hierarchy shown in Fig. 6.2, which has several sub-criteria under each criterion.

This is a more complex AHP hierarchy, with local and global default priorities. Moreover, there are five alternatives in the hierarchy. The local priorities represent the relative weights of the nodes within a group of siblings with respect to their parent. The local priorities of each group of criteria and their sibling sub-criteria add up to 1.000. The global priorities, shown in black, are obtained by multiplying the local priorities of the siblings by their parent's global priority. The global priorities for all the sub-criteria in the level add up to 1.000.

The rule is as follows: Within a hierarchy, the global priorities of child nodes always add up to the global priority of their parent. Within a group of children, the local priorities add up to 1.000.

So far, we have looked only at default priorities. As the AHP moves forward, the priorities will change from their default values as the DMs input information about

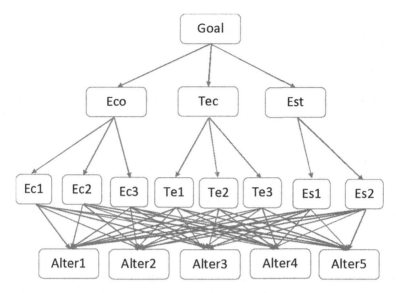

Fig. 6.2 Hierarchy with 3 criteria and 8 sub-criteria (Car hierarchy)

the importance of the various nodes. They do this by making a series of pairwise comparisons.

When economic, social, environmental, and other factors have been reduced to numbers measured in monetary units (CZ crowns, euros, dollars), the weight of objects in tons or grams, and time in seconds, hours, and years have been calculated and probabilities have been estimated, our modeling of complex human problems will often would have reached the limits of its effectiveness. Our mind structures the complex reality into parts and more detailed elements in a hierarchic way. By experiments we have learned that the number of these elements is limited from 5 to 9 (i.e., 72). By decomposition of the system to some homogeneous units and then structuring them hierarchically into some smaller parts, an enormous amount of information can be aggregated into the detailed picture of reality. In the hierarchic structure, a hierarchy is a special type of system based on the assumption that the identified elements of the system can be grouped into the disjoint sets, where the elements of one group influence the elements of the other one, and they are, on the other hand, influenced by elements of the other groups. The elements of the group, called the level or cluster, are mutually independent. The simplest type of hierarchy is the three-level hierarchy, which is, in fact, the classical MCDM problem (see Fig. 6.1).

Example 6.1 The following MCDM problem "Optimal choice of family house" with 3 alternatives: A1, A2, A3
and 6 criteria:
K1—"price"
K2—"age"

K3—"size"
K4—"public transport"
K5—"equipment"
K6—"surroundings"

This is an example of the three-level hierarchic MCDM problem, see Fig. 6.1.

6.3.1   Normalization of Criteria

According to cognitive psychologists there are two types of comparisons: *absolute* and *relative*. In an absolute comparison the alternatives are compared with the given standard, which is evolved on the basis of the DM's experiences in the past. The corresponding criterion according to which each DM evaluates absolutely is called the *cardinal criterion*.

In a relative comparison the alternatives are compared in pairs, usually according to the evaluation terms such as "worse", "better", etc. In the AHP, both types of comparisons are used. Here, we deal in detail with the approach leading to the cardinalization of the pairwise comparisons and then the result is achieved in the form of weights w_{ki}, for each element "i" from the kth hierarchic level L_k, with respect to the higher level L_{k-1}.

In the pairwise comparison, the ith and jth variant from L_k are compared with respect to the joint "criterion". The result of the comparison is interpreted as an estimate of the ratio $\frac{w_{ki}}{w_{kj}}$ on the given scale. In this way the matrix of pairwise comparisons (CM) is obtained. From this matrix the maximal the corresponding priority vector (eigenvector) is calculated and then by normalization the required weights w_{ki} are obtained. Usually, $h = 3$ hierarchical levels are considered, i.e., classical MCDM problem is investigated, however, here we have eventually $h > 3$ levels. This DM situation is more general as it allows for extending classical MCDM problem by considering sub-criteria, sub-sub-criteria, etc.

Definition 6.1 (*Normalized criterion*) Let $f_i \in L_{k-1}$ be a maximizing cardinal criterion on the set of elements L_k, and let $f_i : L_k \rightarrow \mathbf{R}$. Assume that criterion f_i attains only positive values, i.e., $f_i(x_j) > 0$ for all $x_j \in L_k$. For each $f_i \in L_{k-1}$, we define the normalized criterion G_i :

$$G_i(x) = \frac{f_i(x)}{\sum_{x_j \in L_k} f_i(x_j)}, \quad x \in L_k. \tag{6.1}$$

Normalized criteria (6.1) transform the values of the original criterion into the unit scale [0; 1]. From (6.1), it is clear, that for G_i we obtain the basic normalization formula:

$$\sum_{x_j \in L_k} G_i(x_j) = 1. \tag{6.2}$$

6.3.2 Basic Scale

In AHP the pairwise comparisons are applied on the pairs of homogeneous elements, i.e., elements from the same hierarchic level. The basic scale of absolute values and their intensities are listed in the definition in Fig. 6.3. The odd values $1, 3, 5, 7, 9$ denoted in bold font are considered as the main values and the even values in between are considered as supplementary ones.

The above stated scale, also called the Saaty's scale, has not only been verified in many applications, but has been both theoretically and practically compared with many other scales. The numerical values of the evaluation stages explain how many times the "greater" element surpasses the "smaller" one with respect to the common property of the evaluation criterion. On the other hand, the smaller element has the reciprocal (inverse) value with respect to the greater element, see [19, 22, 27].

6.3.3 Calculation of Weights from the Matrix of Pairwise Comparisons

Pairwise comparisons present relative evaluations of elements from the kth hierarchic level L_k with respect to the given element f from the higher level L_{k-1}. The individual pairs are evaluated by the evaluation grades taken from the corresponding scale, in the case of AHP, Saaty's scale. Basic information for calculating the weights of the individual elements $x_i \in L_k$, with respect to the given "criterion" $f \in L_{k-1}$, is supplied by the matrix of pairwise comparisons $S_f = \{s_{ij}\}$.

Values s_{ij} express the ratio of importance of element x_i to the importance of element x_j, with respect to $f \in L_{k-1}$, i.e., the ratio of weights v_i and v_j :

$$s_{ij} = \frac{v_i}{v_j}, \quad x_i, x_j \in L_k, \quad i, j \in \{1, ..., |L_k|\}, \tag{6.3}$$

where $|L_k|$ denotes the number of elements in L_k.

As the weights v_i are not known in advance, (our goal is to set up the weights!), an additional information about numbers s_{ij}, i.e., elements of the nine-point scale 1 to 9, is required, if x_i is more important than x_j, i.e.,

$$s_{ij} \in \{1, 2, 3, 4, 5, 6, 7, 8, 9\}. \tag{6.4}$$

Otherwise, we have

$$s_{ij} = \frac{1}{s_{ji}}. \tag{6.5}$$

If for elements s_{ij} of matrix $S_f = \{s_{ij}\}$ (6.5) holds, then matrix S_f is said to be *reciprocal*. For calculating the weights of the criteria by the AHP method, we have

Evaluation grade	Comparison elements x and y	Explanation
1	x is equally important as y	Both elements are the same
2	x is a little bit strongly prefered than y in the previous situation	The first element is very weakly more important than the second one
3	x is weakly more important than y	The first element is weakly more important than the second one
4	x is a little bit strongly prefered than y in the previous situation	A little bit stroger preference than the previous situation
5	x is strongly more important than y	The first element is strongly more important than the second one
6	x is a little bit strongly prefered than y in the previous situation	A little bit stroger preference than the previous situation
7	x is extremely more important than y	The first element is very strongly more important than the second one
8	x is a little bit strongly prefered than y in the previous situation	A little bit stroger preference than the previous situation
9	x is absolutely more important than y	The first element is absolutely more important than the second one

Fig. 6.3 The fundamental scale

to calculate the eigenvector of the matrix S_f corresponding to the maximal eigenvalue λ_{max} of the pairwise comparisons matrix S_f. By solving a system of $m = |L_k|$ equations with m unknowns, we obtain the vector of weights $w = (w_1, w_2, ..., w_m)$, written in the vector notation as follows:

$$(S_f - \lambda_{max} I)w = 0, \tag{6.6}$$

or, equivalently:

$$S_f w = \lambda_{max} w, \tag{6.7}$$

where λ_{max} is the maximal eigenvalue of matrix S_f and I is the unit matrix, we obtain the eigenvector w, from which we obtain the vector of weights by normalization:

$$v_i = \frac{w_i}{\|w\|}, \quad i \in \{1, ..., m\}. \tag{6.8}$$

Symbol $\|w\|$ denotes the norm of vector w. The properties of the eigenvalues and eigenvectors of reciprocal PCMs have already been investigated in detail in Chap. 2 (see Theorem 2.1). The other important theorem, Wielandt's theorem 2.2, is useful not only from the theoretical point of view, but also from the practical reasons, as it gives a simple computational method for calculating the eigenvector, as well as the corresponding maximal eigenvalue.

Pairwise comparisons represent the relative evaluations of elements from the kth hierarchic level L_k with respect to the given element f from the higher level L_{k-1} expressed by evaluation grades from the fundamental scale, or from a more extended interval $[1; +\infty[$, or other suitable alo-groups, as it was explained in Chap. 3, where non-integer values are allowed. The starting point for that is the pairwise comparisons matrix $S_f = \{s_{ij}\}$. Values s_{ij} express the ratio of importance of element x_i to the importance of element x_j with respect to $f \in L_{k-1}$, i.e., the ratio of weights v_i and v_j, as in (6.3).

6.3.4 Consistency of a PCM

In pairwise comparison matrix $S = \{s_{ij}\}$ in the multiplicative system we investigated in Chap. 2, the value s_{ij} expresses the ratio of importance of element x_i to the importance of element x_j, with respect to $f \in L_{k-1}$, i.e., the ratio of weights vi and vj:

$$s_{ij} = s_{iq}s_{qj} \text{ for all } i, j, q \in \{1, ..., m\}. \tag{6.9}$$

In practical problems for the DM, full consistency is exceptional, and usually the consistency condition (6.9) is often broken for some or sometimes for all $i, j, q \in \{1, ..., m\}$.

On the other hand, a typical example of consistent matrix of pairwise comparisons is a situation, in which the comparison is performed on the basis of some quantitative criterion, and the weights $v_i > 0$, $v_j > 0$ are known in advance, and then for the elements of the matrix of pairwise comparisons s_{ij} it holds that $s_{ij} = \frac{v_i}{v_j}$, for all $i, j \in \{1, ..., m\}$.

From Chap. 2, the inconsistency index is defined for each single element of the decision hierarchy depending on the priority vector calculated by some method presented in Chap. 2. Now, we return to the hierarchy H, introduced earlier, in order to define the inconsistency index for the whole hierarchy with h hierarchic levels. For each element f of the kth hierarchic level L_k, we obtain the matrix S_f of pairwise comparisons of elements from the lower hierarchic level L_{k+1} and the corresponding inconsistency index I_k^f. Now, the problem is, how to understand an inconsistency index I_H of the whole hierarchy, based on individual partial inconsistency indexes I_k^f, for all $k = 1, ..., h - 1, f \in L_k$. We shall deal with this problem at the end of this chapter.

6.4 Synthesis

In this section, we shall deal with the synthesis of partial evaluations of individual inconsistency evaluations of elements in order to gain an aggregated—total evaluation—inconsistency index. Consider hierarchy H with h levels, where $h \geq 2$. Let us fix the level k and investigate the neighboring levels L_k, L_{k+1}, whose elements are denoted as follows:

$$L_k = \{x_1^k, x_2^k, ..., x_{m_k}^k\}, \tag{6.10}$$

$$L_{k+1} = \{x_1^{k+1}, x_2^{k+1}, ..., x_{m_{k+1}}^{k+1}\}. \tag{6.11}$$

From (6.10) and (6.11), it follows that level L_k has m_k elements and level L_{k+1} has m_{k+1} elements.

Definition 6.2 For each element $x \in L_k$, which is a "criterion" for a pairwise comparison of elements of L_{k+1}, we obtain a pairwise comparisons matrix S_x of elements from L_{k+1}. The corresponding eigenvector priority vector (vector of weights) is denoted as:

$$v^k(x) = (v_1^k(x), v_2^k(x), \cdots, v_{m_{k+1}}^k(x)). \tag{6.12}$$

This vector is called the *priority vector* of the kth hierarchic level with respect to element $x \in L_k$.

Definition 6.3 The *priority matrix of the kth hierarchic level B_k* of hierarchy H, with h levels, $h \geq 2, k \in \{1, 2, h - 1\}$, is an $m_{k+1} \times m_{k+1}$ matrix, the rows of which are composed of priority vectors $v^k(x)$ for all elements $x \in L_k$, as

$$\begin{pmatrix} v_1^k(x_1^k) & \cdots & v_1^k(x_{m_k}^k) \\ \vdots & \ddots & \vdots \\ v_{m_{k+1}}^k(x_1^k) & \cdots & v_{m_{k+1}}^k(x_{m_{k+1}}^k) \end{pmatrix}. \tag{6.13}$$

Definition 6.4 Let $1 \leq p < q \leq h - 1$. By the priority vector of the qth hierarchic level with respect to element $x \in L_p$, we define vector $v_p^q(x)$ as

$$v_p^q(x) = B_q B_{q-1} \cdots B_{p+1} v_p(x). \tag{6.14}$$

By Definition 6.4, the priority vector of qth hierarchic level has the same number of elements as has the $(q + 1)$-th hierarchic level, i.e., m_{q+1}. Vector $v_p^q(x)$ determines the relative importance of element in the level L_{q+1} with respect to the criterion on the higher level L_p. It need not be only an immediately higher level. Formula (6.14) is an extension of (6.13), which is valid only for the immediate neighboring higher hierarchic level. Probably the most frequent case is $p = 1, q = h - 1$. Then the highest hierarchic level contains only a single element: g-"Goal", i.e., $L_1 = \{g\}$, of the kth hierarchic level and the lowest hierarchic level L_h consists of the basic hierarchic elements—decision variants (alternatives). Then by (6.14), the priority vector is a "synthetic" vector of weights with respect to the Goal g:

$$v_1^{h-1}(g) = B_{h-1} B_{h-2} \cdots B_2 v^1(g). \tag{6.15}$$

Example 6.2 (*Classical MCDM problem*) Apply the formula (6.15) three-level hierarchy H, i.e., the classical MCDM problem:

$L_1 \quad = \{g\}$-goal,
$L_2 \quad = \{f_1, f_2, \cdots, f_m\}$-criteria,
$L_3 \quad = \{c_1, c_2, \cdots, c_n\}$-variants.

For $p = 1, q = 2$, formula (6.15) is reduced to

$$v_1^2(g) = B_2 v^1(g), \tag{6.16}$$

where $v^1(g) = (w_1, \cdots, w_m)$ are the weights of the criteria with respect to goal g. By $v_i(f_j)$ we denote the weight of variant c_i with respect to the criterion f_j, then by Definition 6.4 we obtain

$$B_2 = \begin{pmatrix} v_1(f_1) & \cdots & v_1(f_m) \\ \vdots & \ddots & \vdots \\ v_m(f_1) & \cdots & v_m(f_m) \end{pmatrix}. \tag{6.17}$$

The vector of weights of variants with respect to the goal:

$$v^2(g) = \begin{pmatrix} \sum_{i=1}^m w_i v_1(f_i) \\ \vdots \\ \sum_{i=1}^m w_i v_m(f_i) \end{pmatrix}. \tag{6.18}$$

In other words, alternatives c_j has the aggregated weight $\sum_{i=1}^m w_i v_j(f_i)$, $j \in \{1, \cdots, n\}$. By these weights the variants can be ranked or the optimal (i.e., compromise) variant can be found. The end of this section is focused on the total inconsistency index of hierarchy H with $h \geq 2$ hierarchic levels. In the previous section, we have introduced the inconsistency index for each element of the hierarchy except the elements from the lowest hierarchic level L_h.

Definition 6.5 Let $1 \leq k \leq h - 1$. The inconsistency index I_x^k of element x from hierarchic level L_k is defined as follows:

For $k = h - 1$ set

$$I_x^{h-1} = I_x \text{ for } x \in L_{h-1}.$$

For $1 \leq k < h - 1$ the inconsistency index is defined as follows:
If I_y^{k+1} is defined for all $y \in L_{k+1}$, then set

$$I_x^k = \max\{I_x, \sum_{y \in L_{k+1}} v_y^k(x) I_y^{k+1}\} \text{ for all } x \in L_k. \qquad (6.19)$$

The *total inconsistency index* I_H of hierarchy H is defined by

$$I_H = I_g^1. \qquad (6.20)$$

The total inconsistency index of the whole hierarchy H is defined as the inconsistency index of element g (goal) from hierarchic level L_1. First, the weighted average from the inconsistency indices of lower level elements is calculated, i.e., $\sum_{y \in L_{k+1}} v_y^k(x) I_y^{k+1}$, and this average is compared to the original inconsistency index I_x, and the final total inconsistency index is the higher of these two numbers.

6.5 Case Study: Optimal Choice of a Passenger Car

In this first case study we deal with the following DM problem: An optimal choice (i.e., purchase) of a new personal car of a lower middle class for the purpose of a certain University X. This realistic example has been solved with the help of the SW module DAME, in Sect. 6.7. At the very beginning, it should be reminded the subjective character of the problem should be remembered. The choice of the new car has been considered for the purposes of University X. For another purpose, e.g., another subject (a family), another hierarchic model, or other criteria may be required, or at least some other relative importance of the criteria.

The table below shows the data for this problem.

Optimal choice of a new personal car of the lower middle class—data:

Alternative	Price	Power	Consumption	Driving	Safety	Design
Skoda Octavia	330	60	7.2	xx	7	*
Ford Focus	350	55	6.7	xxx	7	***
VW Golf	430	75	5.9	x	9	**
Mazda 3	360	60	7.4	xx	8	**
Hyundai i30	290	55	6.9	xxx	6	*

The more x, the worse.
The more *, the better

The decision criteria Price, Power, and Consumption are quantitative, whereas criteria Driving, Safety, and Design have a qualitative nature. The quantitative and qualitative nature of the criteria will require the pairwise comparisons of pairs of variants. The criteria Power, Safety, and Design are maximization criteria. The criteria Price, Consumption, and Driving are minimization ones. The tables presented in the following section demonstrate the documentation of the problem-solving by the output from the DAME module.

6.6 AHP Procedure: Seven Steps in Decision-Making

Any model is always a representation of certain object and/or idea that enables it to be better understood. When modeling, it is important to make clear the most important elements and alternatives in the DM problem to yourself as the user or to other DMs. When constructing the model there is then a smaller probability of overlooking some essential aspects of your decision, and the process of specifying alternatives can inspire the DM to search for new alternatives that are not clear at first sight. The model itself will remain on record for more possible applications or reevaluation in the future, or for when the circumstances change.

In the first part of the chapter, you will continue in the theoretical foundations of the AHP method from the previous chapter and you will learn how to divide the approach to decision-making into seven subsequent (practical) steps so that the DM problem can be effectively resolved, i.e., to set up the compromise variant with the help of a software tool, e.g., Excel or Expert Choice (EC). In the second part of the chapter you will get to resolve the classical case study: *Optimal choice of a passenger car*, formulated at the end of the previous section.

Next, we summarize the main principles of tshe application of AHP in seven steps.

- **Step 1:** Analysis and definition of the decision-making problem.
- **Step 2:** Elimination of non-feasible alternatives.
- **Step 3:** Structuring the hierarchic model.
- **Step 4:** Partial evaluation by the pairwise comparisons.
- **Step 5:** Synthesis of partial evaluations.
- **Step 6:** Verification of the final decision.
- **Step 7:** Documentation of the decisions.

Step 1: Analysis and definition of the decision-making problem

The first step of the DM process includes three parts:

- identification of the problem,
- identification of alternatives and criteria,
- analysis of alternatives.

Identification of the Problem

Identification of the problem belongs to Simon's first stage of DM—looking for appropriate DM situations, see [28]. In practice, this means the identification of DM opportunities and the DM environment. A proper choice of problems to be analyzed for a later DM depends on their importance and complexity.

Identification of the Alternatives and Criteria

The factors influencing the DM problem are called criteria (properties, attributes, characteristics, or views). In Fig. 6.1—DM problem "Choice of optimal location of the company", where six criteria for evaluation of the company are listed:

- K1—"price of company",
- K2—"age of company",
- K3—"size of company",
- K4—"facilities of company",
- K5—"attractiveness of location of company",
- K6—"accessibility of company for customers".

We consider three alternatives: A1, A2, A3, being in the lower part of the hierarchic model, bellow the level of criteria. It is also possible to start with identification of the criteria and then with some of the identification of alternatives, i.e., to use a "top-down" approach, otherwise, we can apply reverse method, i.e., a "bottom-up" approach.

Sometimes, among the criteria and alternatives there exist interdependences, which means that some of the information about criteria enables the definition of alternatives or vice-versa information about alternatives makes it possible to know information about criteria. In such problems, the AHP structure of the problem must be replaced by a more general analytic network process (ANP).

Analysis of Alternatives

How do we identify the decision alternatives? Usually, it is evident from the nature of the DM problem what alternatives should be considered; sometimes, however, it is not so clear. Generating the decision alternatives may be done by brainstorming, professional discussion, and/or research activities. Here, the human brain plays a decisive role. Even a computer with the most perfect software cannot substitute a creative human mind, even though it might be helpful when organizing or structuring and evaluating alternatives and criteria under consideration.

Remember, that the decision model need not contain all the elements that are met during the process of the identification of the model. There is a real danger that the created model created will be extremely complicated and too extensive, so it will become infeasible for practical decision-making. The model should be so large and complex enough that it represents the most important decision interests, and, at the same time, so small that it reacts to all changes of relevant inputs.

Step 2: Elimination of infeasible alternatives

Setting Up the Aspiration Levels
The *aspiration level* of the given criterion represents the minimal requirements for the variant (from the point of view of this criterion). Setting up the aspiration level depends on the knowledge of the DM about the problem and also on the global goal of the decision-making problem. The higher is the aspiration level, the smaller the number of alternatives that will satisfy this level.

Elimination of Alternatives and Aspiration Levels
If the aspiration level is already set up, eliminating an unsatisfactory alternative is a simple matter. For example, if for a criterion K1 price of house, in the "Family house" example, we set up the aspiration level of 0.5 million, then all houses with a price above 0.5 million are not considered. In the case of more criteria and alternatives, it is helpful to use Excel.

Step 3: Structuring hierarchic model

Creating a hierarchic structure will be demonstrated with the example of the "Choice of personal car" from Sect. 6.5.

By the top-down approach the main goal of the decision-making, here, the goal is "Choice of personal car", see Fig. 6.2. On the second hierarchic level, the following criteria are considered:

- Ec—economic,
- Te—technical,
- Es—aesthetical.

These criteria are given by the nature of the problem.

The next hierarchic level represents the more detailed classification of the above three criteria. The economic criterion is divided into sub-criteria:

- Ec1—price,
- Ec2—fuel consumption,
- Ec3—service,
- Ec4—insurance.

The technical criterion is divided into sub-criteria:

- Te1—power of engine,
- Te2—size of luggage compartment,
- Te3—driving comfort and security.

The aesthetic criterion is divided into sub-criteria:

- Es1—design of the car,
- Es2—reliability and image of producer.

The lowest hierarchic level is the level of alternatives:

- A1—Skoda Fabia,
- A2—Fiat Punto,
- A3—VW Polo,
- A4—Renault Clio,
- A5—Opel Corsa.

A heuristic rule for selecting elements into groups is as follows: there is a good reason to compare some elements with respect to some superior element. If there is no good reason for comparing the elements, then either the group of elements is not well defined, or the compared elements do not belong to the same group. More information should then be assembled about the DM problem or a modification of the hierarchic structure is necessary.

Selecting criteria into groups according to individual factors is desirable for at least two reasons. The first reason is the necessity of forming the hierarchic structure reflecting the goal and criteria of the DM problem. The second reason has a more technical character. Selecting criteria into hierarchic groups enables us to comprise a greater number of criteria by a transparent and comprehensible way and reasonable effort.

Now, consider a decision structure with a large number, say m, criteria at the same level. By the pairwise comparisons method, we need to compare the pairs of the elements—criteria. With an increasing number of pairwise comparisons, the attention of the evaluators quickly decreases and the evaluations lose their validity and relevance, i.e., they are more inconsistent. The capacity of the human brain is limited and, by psychological research, the human brain is able to work together with 5–9 elements at the same time. All these reasons lead to preferring a hierarchic structuring of elements into groups.

Step 4: Partial evaluation and pairwise comparisons

In this section, we shall deal with practical questions associated with calculating corresponding priorities—weights of the criteria in practical DM problems. Pairwise comparisons in AHP

In the AHP, the pairs of elements are evaluated with respect to the criteria or properties that are common to them. For instance, when comparing the design of two particular cars, we evaluate whether the first car has a better design than the second one. Moreover, we have a possibility to evaluate whether the first car has a "much better design" than the second one, or whether it has only "a slightly better design" than the second car. In short, we can express an intensity of relation "to have a better design", particularly, on the nine-point Saaty scale.

In the given group of elements of the same hierarchic level, all elements are compared with respect to the superior elements of the next higher level. In this sense we speak about the elements of the inferior group of elements as about the children, and about elements from the superior group of elements as about the parents. Each pair of elements can be compared from two points of view:

- the first element "i" is compared to the second element "j" with respect to the property induced by the parent element, the result is evaluation s_{ij},
- the second element "j" is compared to the first element "i" with respect to the property induced by the parent element, and the result is evaluation s_{ij}.

In practice, only one of these evaluations is carried out, and the second one is given by the reciprocity property as well as the consistency property. For more details, see Chap. 2.

Evaluation is frequently performed by measuring on some scale, e.g., in Czech crowns, euros, liters, seconds, kilowatts, etc. These scales have been forming during the whole history of mankind. They enable inter-human communication, commerce, and everyday life. Social, political, and other factors cannot be, however, measured by physical or economical tools. A human being is able to accept and distinguish a broad amount of feelings and perceptions. This ability allows us to distinguish and to develop relations between individual elements of the DM problem and define elements that have an influence on him and how strong this influence may be. In the AHP the nine-point Saaty's scale has been created for measuring this impact, as described in the previous chapter.

Types of Comparisons: Importance, Preference, or Likelihood

When comparing two elements, we consider three specific situations:

- the first element is more important than the second one,
- the first element is preferred to the second one,
- the first element is more probable (more likely) than the second one.

Each of the abovementioned situations takes place in different conditions.

We speak about the grade of importance if we compare two criteria or evaluation properties. For example, when buying a car the economic criterion "price" could be more important than the technical criterion "power of engine". The grade of preference is applied when comparing the variants according to some criterion, e.g., when comparing variants "VW Polo" with "Skoda-Fabia" according to the criterion "Safety of driving". The grade of likelihood (probability) is suitably applied in situations where uncertain events or elements are compared. For instance, when comparing two interest rates given by a bank, then "interest rate 3%" is more probable than "interest rate 10%".

Some of our evaluations—the results of pairwise comparisons—are based only on our intuition, which depends on our experience and the length of practice. On the other hand, other evaluations are based on qualitative or quantitative data. The AHP allows for a combination of both approaches into one frame by the help of the weights in order to obtain the final evaluation. The practical aspects of this approach are described in Step 5.

Step 5: Synthesis of partial evaluation

Partial evaluations are synthesized (aggregated) into the final (i.e., total) evaluation based on a particular model by the weights. Mathematically, this approach is described in Chap. 2; now, we shall deal with practical aspects of the synthesis of the partial evaluations. The principle of synthesis is represented by a top-down procedure in calculating priorities—positive numbers smaller than 1, being decomposed to the lowest level of the individual variants. The sum of all priorities of all variants must be equal to 1. The variant with the highest priority is considered as the best, i.e., it is a compromise variant of the DM problem. The sum of values of all alternatives must be equal to one, too.

Step 6: Verification of decision

The verification consists in comparing the results attained by the decision model with an intuition of the DM. In the case of a full correspondence with the intuition, then the result is acceptable and applicable to the real decision—the choice of the compromise variant. If, however, the result does not correspond to the intuition, then the model, i.e., criteria and/or alternatives, requires restructuring. Therefore, a return to the previous step should be performed and a new iteration is necessary. Sometimes, after the identification of discrepancies between the model and intuition, the intuition itself needs to be restructured. Here, an analysis of sensitivity may help. It could answer the question: "What happens, if...?".

Analysis of Sensitivity "What if...?"

When will the solution of the DM problem will look like, if the decision criteria or their relative importance change? The answer to this question can be obtained by a new calculation of the synthesis in the previous Step 5. It is clear that repeated calculation is not possible without a computer and a suitable software tool. Luckily, today's managers are equipped with personal computers and corresponding software tools, e.g., Excel.

Subjective Nature of the Model

Someone could object that changing the relative importance of the criteria may be understood as an unwanted manipulation with the results, so that the results would correspond to previously given wanted results. The answer to this question is: "Right, but you should realize that in real DM situations there is usually no true optimal solution. Any solution to the DM problem is always subjective. The point is that you should incorporate into the model as much relevant information as possible, and, on this basis you could formulate the compromise solution."

If you incorporate into the decision model your own evaluation criteria and corresponding variants, then you obtain a solution—a compromise decision, which fulfills your requirements with the maximum satisfaction.

Step 7: Documentation of decisions, software

Documenting all the steps performed in the DM procedure is sometimes even more important than making a particular decision. The documentary material may serve as a justification of the decisions to other people or of a possible revision of the original decisions for the future decisions under changed decision conditions. Generally, it is accepted that the good decisions made in the past may lead to good decisions in the future. Under uncertainty conditions, however, this statement is not always true. Here, it is important that the decisions made in the past are well documented so that the decision-makers can benefit from this material in the future. In fact, this feature is the main advantage of MCDM. Currently, software tools for solving MCDM problems serve well also for the documentation of the solution process.

For this purpose, we have proposed a new Microsoft Excel add-in DAME (Decision Aid Module for Excel) for solving decision-making problems with certain and uncertain data. The software tool called DAME (in a new version: FuzzyDAME), has been developed at the Silesian University in Opava, School of Business Administration in Karvina by Radomir Perzina and Jaroslav Ramik—the author of this text. Compared to other decision support software products, the FuzzyDAME is free, is able to work with scenarios or multiple decision-makers, and can process crisp or fuzzy evaluations of PCMs (see e.g., [8–10, 12, 15, 16]). It is in English, self-guided, allows for easy manipulation of data and utilizes the capabilities of the widespread spreadsheet MS Excel, see [7, 11, 13, 14]. This add-in is used by students in the courses of Decision Analysis for Managers at the School of Business Administration in Karvina, Silesian University in Opava. It can be recommended also for other students, researchers, or small companies.

DAME in the recent version as FuzzyDAME is available at: http://www.opf.slu.cz/kmme/FuzzyDAME/. On this website, there is information of how to download the files, installation, and application instructions of the software tool.

6.7   Case Study: Optimal Choice of a Passenger Car—Continuation from Sect. 6.5

The decision criteria Price, Power, and Consumption are quantitative, whereas the criteria Driving, Safety, and Design have a qualitative nature. Applying DAME, the qualitative nature of the criteria will require the pairwise comparisons of pairs of alternatives. The criteria denoted in red color are minimization ones. The following Figs. 6.4, 6.5, 6.6, and 6.7 present the output from the FuzzyDAME module.

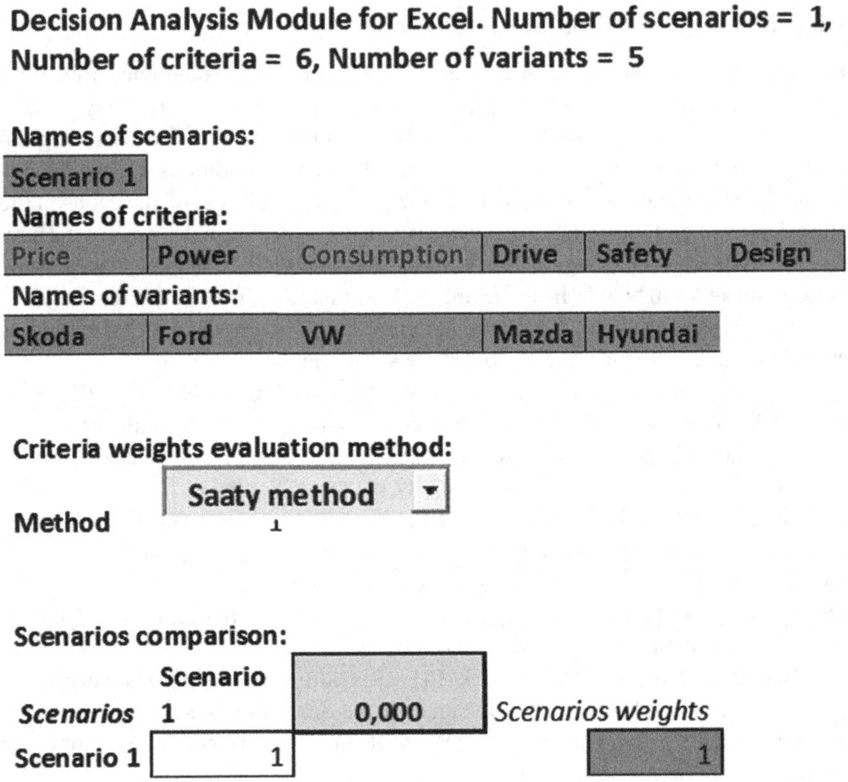

Decision Analysis Module for Excel. Number of scenarios = 1, Number of criteria = 6, Number of variants = 5

Names of scenarios:

Scenario 1

Names of criteria:

Price	Power	Consumption	Drive	Safety	Design

Names of variants:

Skoda	Ford	VW	Mazda	Hyundai

Criteria weights evaluation method:

Method | Saaty method ▼ |
                      ⊥

Scenarios comparison:

	Scenario 1		
Scenarios		0,000	*Scenarios weights*
Scenario 1	1		1

Fig. 6.4 DAME—input data

In Fig. 6.4, input information is summarized. There is a choice of three methods for calculation of the criteria weights: Saaty's eigenvector method, the geometric average method, and direct input of weights.

In Fig. 6.5, the values of all alternatives (variants) according to the quantitative criteria are presented in corresponding tables, as well as all pairwise comparisons matrices for qualitative criteria are presented. In the right upper corner, in a green cell, there is a corresponding inconsistency ratio of the PCM.

In Fig. 6.6, the pairwise comparisons matrix for criteria importance is presented.

The final ranking of alternatives—passenger cars is: 1. Hyundai, 2. VW, 3. Ford, 4. Mazda and 5. Skoda.

Evaluation of Variants According to Individual Criteria:

Price	Value
Skoda	330
Ford	350
VW	290
Mazda	410
Hyundai	390

Power	Value
Skoda	50
Ford	60
VW	80
Mazda	60
Hyundai	70

ꞏnsumptiꞏ	Value
Skoda	72
Ford	67
VW	59
Mazda	74
Hyundai	69

Drive	Skoda	Ford	VW	Mazda	Hyundai	0,045
Skoda	1	1/3	1/2	1	1	
Ford	3	1	3	2	7	
VW	2	0,333333333	1	3	5	
Mazda	1	0,5	0,333333	1	1	
Hyundai	1	0,142857143	0,2	1	1	

Safety	Skoda	Ford	VW	Mazda	Hyundai	0,002
Skoda	1	1	1/2	1/2	1/3	
Ford	1	1	1/2	1/2	1/3	
VW	2	2	1	1	1/2	
Mazda	2	2	1	1	1/2	
Hyundai	3	3	2	2	1	

Design	Skoda	Ford	VW	Mazda	Hyundai	0,047
Skoda	1	1/5	1/3	1/2	1/3	
Ford	5	1	3	4	5	
VW	3	0,333333333	1	1/2	1	
Mazda	2	0,25	2	1	3	
Hyundai	3	0,2	1	0,333333	1	

Fig. 6.5 DAME—results

Criteria Comparison:

Criteria	Price	Power	Consump	Drive	Safety	Design	0,035	Criteria weights
Price	1	7 ▾	2 ▾	3 ▾	1 ▾	6 ▾		
Power	0,142857	1	1/3 ▾	1/5 ▾	1/7 ▾	1/2 ▾		
Consump	0,5	3	1	2 ▾	1/5 ▾	4 ▾		
Drive	0,333333	5	0,5	1	1/3 ▾	1 ▾		
Safety	1	7	5	3	1	5 ▾		
Design	0,166667	2	0,25	1	0,2	1		

Fig. 6.6 DAME—criteria comparison

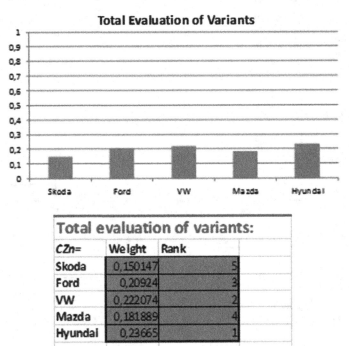

Total evaluation of variants:

CZn=	Weight	Rank
Skoda	0,150147	5
Ford	0,20924	3
VW	0,222074	2
Mazda	0,181889	4
Hyundai	0,23665	1

Fig. 6.7 DAME—final ranking

References

1. Barzilai J (1997) Deriving weights from pairwise comparison matrices. J Oper Res Soc 48(12):1226–1232
2. Bhushan N, Kanwal R (2004) Strategic decision making: applying the analytic hierarchy process. Springer, London
3. Ehrgot M, Figueira JR, Greco S (2014) Trends in multiple criteria decision analysis. Springer International Publishing, Cham
4. Figueira J, Greco S, Ehrgott M (2005) Multiple criteria decision analysis: state of the art surveys. Springer, Berlin

5. Forman EH, Gass SI (2001) The analytical hierarchy process an exposition. Oper Res 49(4):469–487
6. Locatelli G, Mancini M (2012) A framework for the selection of the right nuclear power plant. Int J Prod Res 50(17):4753–4766
7. Ramik J (2007) A decision system using ANP and fuzzy inputs. Int J Innov Comput, Inf Control 3(4):825–837
8. Ramik J, Perzina R (2008) Microsoft excel add-in for solving multicriteria decision problems in fuzzy environment. In: Proceedings of the 26th international conference mathematical methods in economics. Technical University of Liberec
9. Ramik J, Perzina R (2008) Modern metody hodnocen a rozhodovn (Modern methods in evaluation and decision making). Karvina: Silesian University in Opava, School of business administration publishing, 252 p
10. Ramik J, Perzina R (2008) Method for solving fuzzy MCDM problems with dependent criteria. In: Proceedings of the joint 4th international conference on soft computing and intelligent systems and 9th international symposium on advanced intelligent systems. Nagoya University, Nagoya, pp 1323–1328
11. Ramik J, Perzina R (2010) A method for solving fuzzy multicriteria decision problems with dependent criteria. Fuzzy Optim Decis Mak 9(2), 123141
12. Ramik J, Perzina R (2012) DAME microsoft excel add-in for solving multicriteria decision problems with scenarios. In: Proceedings of the 30th international conference mathematical methods in economics. Silesian University, School of Business Administration
13. Ramik J, Vlach M (2013) Measuring consistency and inconsistency of pair comparison systems. Kybernetika 49(3):465–486
14. Ramik J, Perzina R (2014) Solving decision problems with dependent criteria by new fuzzy multicriteria method in excel. J Bus Manag 3:1–16
15. Ramik J, Perzina R (2014) Microsoft excel as a tool for solving multicriteria decision problems. In: Proceedings of the 18th annual conference, KES-2014 Gdynia, Poland. Procedia Computer Science, vol 35. Elsevier, Gdynia, pp. 1455–1463
16. Ramik J, Perzina R (2014) Solving multicriteria decision making problems using microsoft excel. In: Proceedings of the 32nd international conference mathematical methods in economics. Palacky University, Faculty of Science, Olomouc, pp 777–782
17. Saaty TL (1977) A scaling method for priorities in hierarchical structures. J Math Psychol 15(3):234–281
18. Saaty TL (1980) Analytic hierarchy process. McGraw-Hill, New York
19. Saaty TL, Forman EH (1992) The hierarchon: a dictionary of hierarchies. RWS Publications, Pittsburgh
20. Saaty TL (1994) Fundamentals of decision making and priority theory with the AHP. RWS Publications, Pittsburgh
21. Saaty TL, Vargas L (2000) Models, methods, concepts and applications of the analytic hierarchy process. Kluwer, Boston
22. Saaty L (2003) Decision-making with the AHP: why is the principal eigenvector necessary. Eur J Oper Res 145:85–91
23. Saaty TL (2004) Decision making - the analytic hierarchy and network processes (AHP/ANP). J Syst Sci Syst Eng 13(1):1–34
24. Saaty TL (2008) Decision making with the analytic hierarchy process. Int J Serv Sci 1:83–98
25. Saaty TL (2008) Relative measurement and its generalization in decision making: why pairwise comparisons are central in mathematics for the measurement of intangible factors, the analytic hierarchy/network process. RAC SAM Rev R Acad Cien Serie A Mat 102(2), 251–318
26. Saaty TL (2008) Decision making for leaders: the analytic hierarchy process for decisions in a complex world. RWS Publications, Pittsburgh
27. Saaty TL (2010) Principia mathematica decernendi: mathematical principles of decision making. RWS Publications, Pittsburgh
28. Simon H (1960) The new science of management decision. Harper and Brothers, New York
29. Vaidya OS, Kumar S (2006) Analytic hierarchy process: an overview of applications. Eur J Oper Res 169(1):1–29

Chapter 7
Applications in Practical Decision-Making Methods: PROMETHEE and TOPSIS

7.1 Introduction to PROMETHEE

The PROMETHEE (Preference Ranking Organization METHod for Enrichment Evaluation) is a common name for a group of multi-criteria optimization methods, or, optimization methodology which is fundamentally based on pairwise comparisons methods. Historically, it has been evolved in at least in six consecutive versions: PROMETHEE I—PROMETHEE VI.

The PROMETHEE I (partial ranking) and PROMETHEE II (complete ranking) were developed by J. P. Brans and presented for the first time in 1982. In the same year several applications using this methodology were reported already by G. D'Avignon in the field of heath care. A few years later J. P. Brans and B. Mareschal developed PROMETHEE III (ranking based on intervals) and PROMETHEE IV (continuous case). The same authors proposed in 1988 the visual interactive module GAIA, which provides a graphical representation supporting the PROMETHEE methodology. In 1992 and 1994, J. P. Brans and B. Mareschal further suggested two extensions: PROMETHEE V (MCDA including segmentation constraints) and PROMETHEE VI (representation of the human brain). The methods in this group are appropriate for solving various types of problems with all types of information about the criteria and variants. Moreover, there exists a solver—software tool named PROMCALC [4].

A considerable number of applications have made use of the PROMETHEE methodology in various fields such as industrial location, manpower planning, water resources, investments, banking, medicine, chemistry, health care, tourism, etc, see e.g. [2, 7, 12]. The success of the methodology is due to both its mathematical properties and its user-friendliness.

J. Ramik, *Pairwise Comparisons Method*, Lecture Notes in Economics and Mathematical Systems 690, https://doi.org/10.1007/978-3-030-39891-0_7

7.2 Formulation of the Problem

Let us consider the following MCDM problem:

$$\text{"Maximize/Minimize"} \; f_1(c), \, f_2(c), \cdots, f_m(c) \tag{7.1}$$

subject to

$$c \in \mathscr{C} = \{c_1, \cdots, c_n\}, \tag{7.2}$$

where $m, n > 1$ are positive integers, $\mathscr{C}$ is a finite set of *alternatives*, and $f_1, f_2, \cdots, f_m$ is a set of evaluation *criteria*. There is no objection to considering some criteria to be maximized and the others to be minimized. The goal of the DM is to identify an alternative that "optimizes" all the criteria [14, 16].

Usually, this is an ill-posed mathematical problem as in practice there exists no alternative that "optimizes" all the criteria at the same time. However, most (if not all) human problems have a multi-criteria nature. According to our various human aspirations, it makes no sense, and it is often not fair, to select a decision based on a single evaluation criterion only. In most of cases at least technological, economical, environmental, and social criteria should always be taken into account. Multi-criteria problems are, therefore, extremely important and require an appropriate treatment.

The basic data of a multi-criteria problem (7.1), (7.2) consist of a *decision matrix*:

$$D = \begin{pmatrix} f_1(c_1) & f_2(c_1) & \cdots & f_m(c_1) \\ f_1(c_2) & f_2(c_2) & \cdots & f_m(c_2) \\ \vdots & \vdots & \cdots & \vdots \\ f_1(c_n) & f_2(c_n) & \cdots & f_m(c_n) \end{pmatrix}. \tag{7.3}$$

As there is no alternative that "optimizes" all the criteria at the same time, a *compromise alternative* should be selected. Most decision problems have such a multi-criteria nature.

Associated with a multi-criteria problem (7.3) we define *dominance relations* P, I and R, defined as follows: For each $x, y \in \mathscr{C}$:

$$x P y \quad if \quad \forall j : f_j(x) \geq f_j(y) \text{ and } \exists k : f_j(x) > g_j(y). \tag{7.4}$$

$$x I y \quad if \quad \forall j : f_j(x) = f_j(y). \tag{7.5}$$

$$x R y \quad if \quad \exists r : f_r(x) < f_r(y) \text{ and } \exists s : f_s(x) > g_s(y). \tag{7.6}$$

where P, I, and R, respectively, stand for *preference, indifference,* and *incomparability*. This definition is quite obvious. One alternative is better than another if it is at least as good as the other according to all criteria. If one alternative "y" is better on a criterion f_r and the other one, "x", is better on criterion f_s, then it is impossible to decide which alternative is the best one without additional information. Both alternatives are therefore incomparable.

Alternatives that are not dominated by any other alternative are called *efficient alternatives*, or, *non-dominated alternatives*. Given a decision matrix (7.3) for a particular multi-criteria problem, most of the alternatives (often all of them) are usually efficient. The dominance relation is very poor on P and I. When one alternative is better on one criterion, the other is often better on the other criterion. Consequently incomparability holds for most pairwise comparisons, so that it is impossible to decide optimally without additional information.

7.3 Preference Functions

The methods in this group are appropriate for solving various types of problems with all types of information about the criteria and alternatives. The PROMETHEE-type methods are based on pairwise comparisons of alternatives $c_i, c_j \in \mathscr{C}, i, j \in \{1, \cdots, n\}$, by all criteria $f_k, k \in \{1, \cdots, m\}$. We assume, that the criteria are cardinal and their relative importance is given by the weights w_k. We can use any evaluation system, not only multiplicative, but also additive, fuzzy, or any other one, in order to obtain the corresponding priority vector suitable for the given DM problem (see Chap. 2). The result of pairwise comparisons of $c_i, c_j \in \mathscr{C}$, by f_k is a number $P_k(c_i, c_j) \in S$, where usually, $S = [0; 1]$, and the intensity depends on the difference of values of the criterion $d_k = f_k(c_i) - f_k(c_j)$. If criterion f_k is maximizing (i.e., the higher the value, the better), then the greater the difference, the higher the preference. A particular preference is given by the preference function $Q : \mathbf{R} \to [0; 1]$, where

$$P_k(c_i, c_j) = Q(f_k(c_i) - f_k(c_j)) = Q(d_k). \tag{7.7}$$

The PROMETHEE-type methods utilize six basic types of preference functions $Q_i, i = 1, 2, \ldots, 6$. Moreover, for each preference function Q_i there is an associated *threshold of preference* p^*, *threshold of indifference* q^*, and *standard deviation* σ^* is associated. Now, we shall present all six types of the preference function Q_i.

Preference Function Q_1

By the preference function Q_1 we define the result of pairwise comparisons as follows: the preference value is equal to 1, whenever the values of the criteria are different. The value of the preference function is 0 if and only if the values of the criteria are equal, i.e.,

$$\begin{aligned} Q_1(d_k) &= 0 \quad \text{if } d_k = 0, \\ &= 1 \quad \text{otherwise.} \end{aligned} \tag{7.8}$$

See Fig. 7.1.

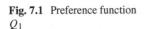

Fig. 7.1 Preference function Q_1

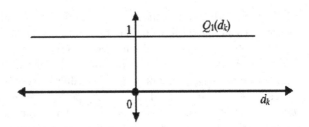

Fig. 7.2 Preference function Q_2

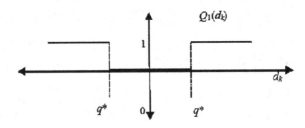

Preference Function Q_2

The preference function Q_2 defines the result of pairwise comparisons similarly to Q_1, except for the range defined by the threshold of indifference, q^*, where the difference of the values of the criteria is considered as indifferent, i.e.,

$$Q_2(d_k) = 0, \quad \text{if } |d_k| \leq q^*,$$
$$= 1 \quad \text{otherwise.} \tag{7.9}$$

See Fig. 7.2.

Preference Function Q_3

The preference function Q_3 enables the grade of preference also to obtain the other values from the interval $[0; 1]$. The user sets up the threshold of preference p^*. If the difference of values of the criteria is less than the threshold, then the grade of preference is less than 1 and linearly decreases down to the zero value, i.e.,

$$Q_3(d_k) = \frac{|d_k|}{q^*}, \quad \text{if } |d_k| \leq p^*,$$
$$= 1 \quad \text{otherwise.} \tag{7.10}$$

See Fig. 7.3.

Preference Function Q_4

The preference function Q_4 defines the result of pairwise comparisons by 3 values: 0 and 1 and now also 0,5. The user sets up the threshold of preference p^* and also the threshold of indifference q^*, which must be smaller than the threshold of

Fig. 7.3 Preference function
Q_3

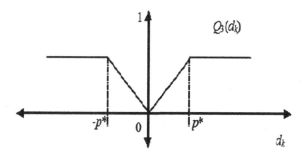

Fig. 7.4 Preference function
Q_4

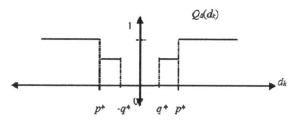

preference. If the difference of values of the criteria is between these thresholds, then the preference function is set to 0.5, i.e.,

$$Q_4(d_k) = 0, \ \text{if } |d_k| \leq q^*,$$
$$= 0,5 \ \text{if } q^* < |d_k| \leq p^*, \tag{7.11}$$
$$= 1 \ \text{otherwise.}$$

See Fig. 7.4.

Preference Function Q_5

The preference function Q_5 combines the properties of the previous two preference functions, i.e.,

$$Q_5(d_k) = 0, \ \text{if } |d_k| \leq q^*,$$
$$= \frac{|d_k|-q^*}{p^*-q^*}, \ \text{if } q^* < |d_k| \leq p^*, \tag{7.12}$$
$$= 1 \ \text{otherwise.}$$

See Fig. 7.5.

Preference Function Q_6

The preference function Q_6 has a special form transforming the difference in evaluation by Gauss's function, the well-known shape of the probability function. It is necessary to set up the parameter σ, a standard deviation of values of the criterion. The value of the preference function tends to be 1 in limit when the difference of evaluation is increasing, i.e.,

Fig. 7.5 Preference function
Q_5

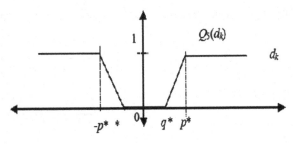

Fig. 7.6 Preference function
Q_6

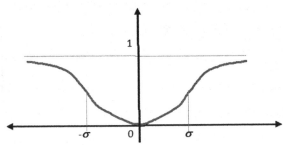

$$Q_6(d_k) = 1 - e^{\frac{-d_k^2}{2\sigma^{*2}}}. \qquad (7.13)$$

See Fig. 7.6.

Now, we assume, that for each pair of alternatives c_i, c_j and each criterion f_k with the help of the chosen preference function Q_k and (7.8), the value of preference $P_k(c_i, c_j)$ is calculated as the number from the interval $[0; 1]$. Notice that Q_k may be arbitrarily chosen from the above six specific preference functions.

We define the *global preference index* $P(c_i, c_j)$ of alternatives c_i, c_j as the weighted average of the preference values $P_k(c_i, c_j)$ and weights w_k as

$$P(c_i, c_j) = \sum_{k=1}^{m} w_k P_k(c_i, c_j), \qquad (7.14)$$

where w_k are the corresponding weights. The *positive flow* F_i^+ of alternative c_i, resp., *negative flow* F_i^- of variant c_i, we call the arithmetic average of the global preference indices $P(c_i, c_j)$, resp. $P(c_j, c_i)$:

$$F_i^+ = \frac{1}{m} \sum_{j=1}^{m} P(c_i, c_j), \text{ resp. } F_i^- = \frac{1}{m} \sum_{j=1}^{m} P(c_j, c_i), \qquad (7.15)$$

The *flow* F_i of alternative c_i we call the difference between the positive and negative flows of alternative c_i, i.e.,

$$F_i = F_i^+ - F_i^-. \qquad (7.16)$$

Here, we shall deal briefly only with the PROMETHEE I and PROMETHEE II methods. Other types of PROMETHEE methods can be found in the literature (see e.g., [1, 5, 6]).

The method of PROMETHEE I defines the partial ordering of the alternatives on the basis of the values of positive and negative flows:

- Alternative c_i is preferred to alternative c_j if $F_i^+ \geq F_j^+$ and $F_i^- \leq F_j^-$. At least one of the inequalities must be strict.
- Alternative c_i is indifferent to alternative c_j if $F_i^+ = F_j^+$ and $F_i^- = F_j^-$.
- Otherwise, the alternatives c_i and c_j are considered as incomparable.
- The approach of transitive closure can be applied to the resulting relation of partial ordering by removing the cycles (see [6]).

In the method of PROMETHEE II the total ordering of alternatives (and also obtaining the compromise alternative) is obtained by decreasing the value of the flow of alternatives F_i.

7.4   Case Study: Optimal Choice of Personal Computer

We consider an MCDM problem with 5 alternatives: $c_1 = $ PC1, $c_2 = $ PC2, $c_3 = $ PC3, $c_4 = $ PC4, $c_5 = $ PC5 (computers PC) and 4 criteria f_1—price, f_2—processor, f_3—HD, f_4—multimedia, as shown in the following table (Table 7.1):

Order the alternatives from the best to the worst by the PROMETHEE II method.

Solution:

First, we choose for each criterion a suitable preference function and the corresponding parameters—the thresholds of preference and indifference. For criterion f_1—price we take the preference function Q_5, where a difference in the price less than 1.5 is considered as insignificant (i.e., $q^* = 1.5$) and a difference in price greater than 8, i.e., $p^* = 8$, is absolutely significant. For criterion f_2—processor and f_4—multimedia, we apply the preference function Q_2, where a difference smaller than (resp. greater than) 2, i.e., $p^* = q^* = 2$, is considered nonsignificant (resp. absolutely significant). For criterion f_3—HD, the preference function Q_1 is used, as this criterion attains only 3 values and each difference is absolutely significant. For each

Table 7.1 Data: 4 criteria, 5 alternatives

Alternative/Crit.	f_1	f_2	f_3	f_4
PC1	15.5	2	40	3
PC2	12	3.5	20	2
PC3	3.5	40	40	4
PC4	5	20	20	5
PC5	1	55	80	1
Weights	0.143	0.264	0.318	0.275

Table 7.2 Positive and negative flows

$P(c_i, c_j)$	PC1	PC2	PC3	PC4	PC5	F^+
PC1	–	0.095	0.064	0.115	0.135	0.102
PC2	0.041	–	0.021	0.084	0.135	0.070
PC3	0.021	0.031	–	0.095	0.135	0.070
PC4	0.104	0.031	0.052	–	0.135	0.080
PC5	0.052	0.095	0.115	0.115	–	0.094
F^-	0.054	0.063	0.063	0.102	0.135	–

Table 7.3 Final rank of alternatives

c_i	F	Rank
PC1	0.0477	1
PC2	0.0075	2
PC3	0.0074	3
PC4	−0.0221	4
PC5	−0.0405	5

pair of alternatives, the global preference index $P(c_i, c_j)$ is calculated by (7.14), then the positive and negative flows by (7.16), as in the following table (Table 7.2):

Now, we are reminded to calculate for each variant the values of flows by (7.16) and then order the alternatives (Table 7.3).

Hence, by PROMETHEE II, the optimal (compromise) alternative is PC1.

7.5 Introduction to TOPSIS Method

The Technique for Order of Preference by Similarity to Ideal Solution (TOPSIS) is a multi-criteria decision analysis method, which was originally developed by Ching-Lai Hwang and Yoon [9] with further developments by Yoon [15] and Hwang, Lai and Liu [10]. TOPSIS is based on the concept that the chosen alternative should have the shortest "distance" from the positive ideal alternative (PIA) and the longest "distance" from the negative ideal alternative (NIA). An important part of TOPSIS model is based on pairwise comparisons.

Let us consider the following multi-criteria problem:

$$\text{"Maximize" } f_1(c), f_2(c), …, f_m(c) \qquad (7.17)$$

subject to

$$c \in \mathscr{C} = \{c_1, …, c_n\}, \qquad (7.18)$$

where $m, n > 1$ are positive integers, $\mathscr{C}$ is a finite set of *alternatives* and

$$f_1(c), f_2(c), …, f_m(c)$$

is a set of *evaluation criteria*. The goal of the DM is to identify an alternative optimizing all the criteria. The basic data of a multi-criteria problem (7.17), (7.18) consist of a *decision matrix*:

$$D = \begin{pmatrix} f_1(c_1) & f_2(c_1) & \cdots & f_m(c_1) \\ f_1(c_2) & f_2(c_2) & \cdots & f_m(c_2) \\ \vdots & \vdots & \cdots & \vdots \\ f_1(c_n) & f_2(c_n) & \cdots & f_m(c_n) \end{pmatrix}. \tag{7.19}$$

As there is no alternative that "optimizes" all the criteria at the same time, a *compromise alternative* can be selected. Here, the TOPSIS method is proposed. Problems (7.17), (7.18) will be transformed into another problem that the "optimal" alternative should have the shortest "distance" from the positive ideal solution (PIS) and the longest "distance" from the negative ideal solution (NIS), see below. including the solution by TOPSIS method is TOPSIS is a method of compensatory aggregation that compares a set of alternatives by identifying *weights* for each criterion, normalizing scores for each criterion and calculating the geometric distance between each alternative and the ideal alternative, applying a suitable distance function (Euclidean, Minkowski's, or Chebyshev's distance function). An assumption of TOPSIS is that the criteria are monotonically increasing or decreasing. Normalization is usually required as the parameters or criteria are often of incongruous dimensions, or antagonistic in multi-criteria problems [9, 13]. Compensatory methods such as TOPSIS allow trade-offs between criteria, where a poor result in one criterion can be negated by a good result in another criterion. This provides a more realistic form of modeling than non-compensatory methods, which include or exclude alternative solutions based on hard cut-offs [11]. An example of application on nuclear power plants is below. A more realistic case study of a similar problem is provided in [8].

7.6 Description of the TOPSIS Method

The Technique for Order of Preference by Similarity to the Ideal Solution (TOPSIS) is based on the concept of the distance function usually based on the concept of a norm. The norm $\| \cdot \|$ on $\mathbf{R}^N$ (N is a positive integer) is a nonnegative-valued function $\| \cdot \| : \mathbf{R}^N \to [0, +\infty[$ with the following properties [3]:

For all $\alpha \in \mathbf{R}$ and all $u, v \in \mathbf{R}^N$

(i)   If $\|v\| = 0$ then $v = 0$,
(ii) $\|\alpha v\| = |\alpha| \|v\|$,
(iii) $\|u + v\| \leq \|u\| + \|v\|$.

Function $d : \mathbf{R}^N \times \mathbf{R}^N \to \mathbf{R}$ is called the *distance function in $\mathbf{R}^N$* (*metric function in $\mathbf{R}^N$*), if the following assumptions are satisfied:

(i) $d(x, y) \geq 0$ for all $x, y \in \mathbf{R}^N$, ("nonnegativity"),
(ii) $d(x, y) = d(y, x)$ for all $x, y \in \mathbf{R}^N$, ("symmetry"),
(iii) $d(x, y) = 0$ if and only if $x = y$, ("uniqueness"),

(iv) $d(x, z) \leq d(x, y) + d(y, z)$ for all $x, y, z \in \mathbf{R}^N$, ("triangle inequality").

Clearly, the function $d : \mathbf{R}^N \times \mathbf{R}^N \to \mathbf{R}$ defined for all $x, y \in \mathbf{R}^N$ by

$$d(x, y) = \|x - y\| \tag{7.20}$$

is a distance function.

Specifically, we are interested the parametrized function

$$d(x, y) = (\sum_{i=1}^{N}(x_i - y_i)^p)^{\frac{1}{p}}, \tag{7.21}$$

where $x = (x_1, ..., x_N)$, $y = (y_1, ..., y_N)$, and $p > 0$ is a parameter. It is clear that function (7.21) satisfies conditions (i)–(iii). The proof of (iv) is more complicated and can be found in [3].

For some values of the parameter $p \geq 1$, the distance function (7.21) is known under special names. For $p = 2$, the distance function (7.21) is called the *Euclidean distance* and for each $x = (x_1, ..., x_N)$, $y = (y_1, ..., y_N)$ has the form:

$$d_E(x, y) = (\sum_{i=1}^{N}(x_i - y_i)^2)^{\frac{1}{2}}. \tag{7.22}$$

For $p \to +\infty$, the distance function (7.21) is called the *Chebyshev's distance* and has the form:

$$d_C(x, y) = \max_{i=1,...,N}\{|x_i - y_i|\}. \tag{7.23}$$

For $p = 1$, the distance function (7.21) is called the *Minkowski's distance* (or *Manhattan's distance*) and has the form:

$$d_M(x, y) = \sum_{i=1}^{N}|x_i - y_i|. \tag{7.24}$$

The Euclidean distance function is the most popular in practice.

Let us consider the decision matrix D from (7.19), which consists of alternatives and criteria described by

$$D = \begin{pmatrix} x_{11} & x_{12} & \cdots & x_{1m} \\ x_{21} & x_{22} & \cdots & x_{2m} \\ \vdots & \vdots & \cdots & \vdots \\ x_{n1} & x_{n2} & \cdots & x_{nm} \end{pmatrix}, \tag{7.25}$$

where $x_{ij} = f_j(c_i), i \in \{1, \cdots, n\}, j \in \{1, \cdots, m\}$ indicates the evaluation of the alternative c_i according to the criterion f_j. The positive weight vector $w =$

$(w_1, \cdots, w_n)$ is composed of the individual weights w_j, $j \in \{1, \cdots, m\}$, for each criterion f_j satisfying $\sum_{j=1,\cdots,m} w_j = 1$.

The data of the decision matrix D come from different sources, so it is necessary to normalize it in order to transform it into a dimensionless matrix, which allows the comparison of the various criteria. Here, we normalize the values by the suitable norm of the corresponding vector of values of the criterion. Then we obtain the *normalized decision matrix* D_{norm}, i.e.,

$$D_{norm} = \begin{pmatrix} y_{11} & y_{12} & \cdots & y_{1m} \\ y_{21} & y_{22} & \cdots & y_{2m} \\ \vdots & \vdots & \cdots & \vdots \\ y_{n1} & y_{n2} & \cdots & y_{nm} \end{pmatrix}, \tag{7.26}$$

where

$$y_{ij} = \frac{x_{ij}}{\|x_j\|}, \tag{7.27}$$

and $x_j = (x_{1j}, \cdots, x_{nj})$, $i \in \{1, \cdots, n\}$, $j \in \{1, \cdots, m\}$. Any of the norms (7.22)–(7.24) can be applied depending on the particular problem.

The normalized decision matrix D_{norm} represents the relative rating of the alternatives. After normalization, we calculate the *weighted normalized decision matrix*, D_w, composed of the matrix D_{norm} and vector of weights $w = (w_1, \cdots, w_n)$, evaluated by some pairwise comparisons method, as follows:

$$D_w = \begin{pmatrix} z_{11} & z_{12} & \cdots & z_{1m} \\ z_{21} & z_{22} & \cdots & z_{2m} \\ \vdots & \vdots & \cdots & \vdots \\ z_{n1} & z_{n2} & \cdots & z_{nm} \end{pmatrix}, \tag{7.28}$$

where

$$z_{ij} = y_{ij} w_j, \quad j \in \{1, \cdots, m\}, \quad i \in \tag{7.29}$$

and $z_j = (z_{1j}, \cdots, x_{nj})$, $i \in \{1, \cdots, n\}$, $j \in \{1, \cdots, m\}$.

The criteria are classified into two types: the benefit criteria (J_B) and cost ones (J_C), i.e., $J_B \cup J_C = \{1, \cdots, m\}$. The benefit criterion means that the higher value the better (from the point of view of the decision-maker, while for the cost criterion is valid the opposite. We define the *positive ideal alternative* z^+ (*benefits*) and *negative ideal alternative* z^- (*costs*) as

$$z^+ = (z_1^+, z_2^+, \cdots, z_n^+), \quad z^- = (z_1^-, z_2^-, \cdots, z_n^-), \tag{7.30}$$

where

$$z_j^+ = \max\{z_{ij} | i \in J_B\}, \quad z_j^+ = \min\{z_{ij} | i \in J_C\}, \tag{7.31}$$

and

$$z_j^- = \min\{z_{ij} | i \in J_B\}, \quad z_j^- = \max\{z_{ij} | i \in J_C\}. \tag{7.32}$$

7.7 The Algorithm

The TOPSIS method is described in the following five steps [8]:

Step 1:
Identify the alternatives $c_1, \cdots, c_n$ and criteria $f_1, \cdots, f_m$ of the MCDM problem, then calculate the decision matrix D by (7.25). Evaluate the weights $w = (w_1, \cdots, m)$ of the criteria by a pairwise comparison method.

Choose the appropriate norm $\| \cdot \|$ and corresponding distance function $d(\cdot)$.

Step 2:
Calculate the normalized matrix D_{norm} by (7.26) and (7.27). Then calculate the positive ideal alternative z^+ (benefits) and negative ideal alternative z^- (costs) by (7.31) and (7.32) as follows:

$$z^+ = (z_1^+, z_2^+, \cdots, z_m^+),$$

$$z^- = (z_1^-, z_2^-, \cdots, z_m^-).$$

Step 3:
Calculate the corresponding distances from the positive ideal alternative z^+ and the negative ideal alternative z^- of each alternative $z_i, i \in \{1, \cdots, n\}$, respectively, as follows:

$$d_i^+ = d(z_i, z^+) = \|z_i - z^+\|, i = 1, \cdots, m, \quad d_i^- = d(z_i, z^-) = \|z_i - z^-\|, i = 1, \cdots, m. \tag{7.33}$$

Step 4:
Calculate the relative closeness D_i^* as follows:

$$D_i^* = \frac{d_i^-}{d_i^+ + d_i^-}, \quad i = 1, \cdots, m. \tag{7.34}$$

Step 5:
Rank the alternatives c_i according to the relative closeness $D_i^*, i = 1, \cdots, m$. The best alternatives are those that have a higher value of D_i^*, because they are closer to the positive ideal alternative and/or farther to the negative ideal alternative.

Remark 7.1 Evidently, $0 \leq D_i^* \leq 1$, as by (7.34) $D_i^* = \frac{1}{\frac{d_i^+}{d_i^-}+1}$, and $d_i^+ \geq 0$, $d_i^- \geq 0$.

7.8 Application of TOPSIS: An Example

Consider six criteria in the problem of optimal choice of a suitable location of anticipated nuclear power plant in preselected localities L1, L2, L3:

f_1—number of workers (cost criterion),
f_2—performance output in MW (benefit criterion),
f_3—total investment costs in billions of CzK (cost criterion),
f_4—operation costs in millions of CzK /year (cost criterion),
f_5—number of evacuated residents (cost criterion),
f_6—grade score of reliability of operation (benefit criterion).

In the following table, you can see the input data for this MCDM problem decision matrix H. In the last row there are the weights of the criteria (Table 7.4):

Solution:

We apply the 5-step algorithm of TOPSIS method from Sect. 7.7 for three types of the norm: Euclidean, Minkovski's, and Chebyshev's norm. In Fig. 7.7, the pairwise comparisons matrix for criteria importance is presented.

In Fig. 7.7, the original data and the normalized data are depicted in Fig. 7.8.

In Fig. 7.9, the final results for corresponding distances as well as the final rankings of alternatives based on the relative closeness are summarized.

Table 7.4 Location of power stations: Data

Locations	f_1	f_2	f_3	f_4	f_5	f_6
L1	6500	4000	90	400	5500	9
L2	4800	2400	50	300	4000	7
L3	7500	4800	100	400	6000	8
Weights	0.07	0.24	0.33	0.09	0.08	–

Data	f1 -	f2 +	f3 -	f4 -	f5 -	f6 +
L1	6500	4000	90	400	5500	9
L2	4800	2400	50	300	4000	7
L3	7500	4800	100	400	6000	8
x+	4800	4800	50	300	4000	9
x-	7500	2400	100	400	6000	7
w	0,07	0,24	0,33	0,19	0,09	0,08
Eukl.norm	11024,52	6693,28	143,53	640,31	9069,18	13,93
Mink.norm	18800	11200	240	1100	15500	24
Cheb.norm	7500	4800	100	400	6000	9

Fig. 7.7 Original data and the normalization coefficients

Normalized data:

Eukl.norm	f1 -	f2 +	f3 -	f4 -	f5 -	f6 +
L1	0,590	0,598	0,627	0,625	0,606	0,646
L2	0,435	0,359	0,348	0,469	0,441	0,503
L3	0,680	0,717	0,697	0,625	0,662	0,574
x+	0,435	0,717	0,348	0,469	0,441	0,646
x-	0,680	0,218	0,697	0,625	0,662	0,503

Mink.norm	f1 -	f2 +	f3 -	f4 -	f5 -	f6 +
L1	0,346	0,357	0,375	0,364	0,355	0,375
L2	0,255	0,214	0,208	0,273	0,258	0,292
L3	0,399	0,429	0,417	0,364	0,387	0,333
x+	0,255	0,429	0,208	0,273	0,258	0,375
x-	0,399	0,214	0,417	0,364	0,387	0,292

Cheb.norm	f1 -	f2 +	f3 -	f4 -	f5 -	f6 +
L1	0,867	0,833	0,900	1,000	0,917	1,000
L2	0,640	0,500	0,500	0,750	0,667	0,778
L3	1,000	1,000	1,000	1,000	1,000	0,889
x+	0,640	1,000	0,500	0,750	0,667	1,000
x-	1,000	0,500	1,000	1,000	1,000	0,778

Fig. 7.8 Normalized data for three types of norm

Eukl.*w.	f1 -	f2 +	f3 -	f4 -	f5 -	f6 +	d+	d-	D=d-/(d+ +d-)	Rank
L1	0,041	0,143	0,207	0,119	0,055	0,052	0,102	0,095	0,481	3
L2	0,030	0,086	0,115	0,089	0,040	0,040	0,087	0,126	0,592	1
L3	0,048	0,172	0,230	0,119	0,060	0,046	0,122	0,120	0,496	2
x+	0,030	0,172	0,115	0,089	0,040	0,052				
x-	0,048	0,052	0,230	0,119	0,060	0,040				

Mink.*w	f1 -	f2 +	f3 -	f4 -	f5 -	f6 +	d+	d-	D=d-/(d+ +d-)	Rank
L1	0,024	0,086	0,124	0,069	0,032	0,030	0,507	0,452	0,471	3
L2	0,018	0,051	0,069	0,052	0,023	0,023	0,359	0,328	0,477	1
L3	0,028	0,103	0,138	0,069	0,035	0,027	0,540	0,484	0,472	2
x+	0,018	0,103	0,069	0,052	0,023	0,030				
x-	0,028	0,051	0,138	0,069	0,035	0,023				

Cheb.*w	f1 -	f2 +	f3 -	f4 -	f5 -	f6 +	d+	d-	D=d-/(d+ +d-)	Rank
L1	0,061	0,200	0,297	0,190	0,083	0,080	0,502	0,476	0,487	2
L2	0,045	0,120	0,165	0,143	0,060	0,062	0,347	0,308	0,471	3
L3	0,070	0,240	0,330	0,190	0,090	0,071	0,534	0,510	0,488	1
x+	0,045	0,240	0,165	0,143	0,060	0,080				
x-	0,070	0,120	0,330	0,190	0,090	0,062				

Fig. 7.9 Final results and rankings for three types of norm

7.9 Conclusion of Applications of PCMs in Practical Decision-Making Problems

The last two chapters of this monograph are focused on the practical application of pairwise comparisons in the three most popular decision-making methods: AHP, PROMETHEE, and TOPSIS. All the methods are briefly introduced and described with a special emphasis on the features and functioning of pairwise comparisons. From this point of view, all the methods are revisited.

The analytic hierarchy process (AHP) is a theory of relative measurement on absolute scales of both quantitative and qualitative criteria based on paired comparison judgment of knowledgeable experts. Comparisons can serve as a tool of measurement and enable a valid scale to be derived from these measurements. How to measure intangibles is the main concern of the mathematics of the AHP. But it must work with tangibles as well to give back measurements that will be used for the tangible factors in a decision with accuracy on tangibles. A fundamental problem of decision theory is how to derive weights for a set of activities according to their importance. Importance is usually judged according to several criteria. Each criterion may be shared by some or by all of the activities. The criteria may, for example, be objectives that the activities have been devised to fulfill. This is a process of multiple criteria decision-making which we study here through a theory of measurement in a hierarchical structure. The chapter is closed with a case study demonstrating the DAME, a special original software tool (Excel add-in) for solving MCDM problems.

The methods of the type PROMETHEE are based on pairwise comparisons of n alternatives by m criteria. We assume, that the criteria are cardinal and their relative importance is given by the weights. We can use any evaluation system, not only multiplicative, but also additive, fuzzy, or any other one, in order to obtain the corresponding priority vector suitable for the given DM problem. The result of pairwise comparisons of alternatives is a number from the interval [0; 1], and the intensity depends on the difference of values of the criterion. If the criterion is maximizing (i.e., the higher the value, the better), then the greater the difference, the higher the preference. A particular preference is given by the preference functions. The methods of type PROMETHEE utilize six basic types of preference functions; moreover, for each preference function, there is an associated threshold of preference, threshold of indifference, standard deviation. An illustrative case study demonstrating the PROMETHEE II method is presented and discussed.

The Technique for Order of Preference by Similarity to Ideal Solution (TOPSIS) is a multi-criteria decision analysis method, which is based on the concept that the "optimal" alternative should have the shortest "distance" from the positive ideal alternative (PIA) and the longest "distance" from the negative ideal alternative (NIA). An important part of TOPSIS model is based on pairwise comparisons. Here, we present a generalized version of the TOPSIS method known from the literature.

References

1. Behzadian M, Kazemzadeh RB, Albadvi A, Aghdasi AM (2010) PROMETHEE: a comprehensive literature review on methodologies and applications. Eur J Oper Res 200:198–215
2. Behzadian M, Otaghsara SK, Yazdani M, Ignatius J (2012) A state-of the-art survey of TOPSIS applications. Expert Syst Appl 39:13051–13069
3. Boyd S, Vandenberghe L (2004) Convex optimization. Cambridge University Press, Cambridge
4. Brans JP, Marechal B (1994) The PROMCALC and GAIA decision support system for multi-criteria decision aid. Decis Support Syst 12(5):297–310
5. Brans JP, Vincke P (1985) A preference ranking organisation method: the PROMETHEE method for MCDM. Manag Sci 31(6):647–656
6. Figueira J, Greco S, Ehrgott M (2005) Multiple criteria decision analysis: state of the art surveys. Springer, Berlin
7. Huang IB, Keisler J, Linkov I (2011) Multi-criteria decision analysis in environmental science: ten years of applications and trends. Sci Total Environ 409(19):3578–3594
8. Huang J (2008) Combining entropy weight and TOPSIS method for information system selection. In: Proceedings of the IEEE conference on cybernetics and intelligent systems, CIS'2008, pp 1281–1284
9. Hwang CL, Yoon K (1981) Multiple attribute decision making: methods and applications. Springer, New York
10. Hwang CL, Lai YJ, Liu TY (1993) A new approach for multiple objective decision making. Comput Oper Res 20(8):889–899
11. Krohling RA, Campanharo VC (2011) Fuzzy TOPSIS for group decision making: a case study for accidents with oil spill in the sea. Expert Syst Appl 38:4190–4197
12. Locatelli G, Mancini M (2012) A framework for the selection of the right nuclear power plant. Int J Prod Res 50(17):4753–4766
13. Lourenzutti R, Krohling RA (2014) The Hellinger distance in multicriteria decision making: an illustration to the TOPSIS and TODIM methods. Expert Syst Appl 41:4414–4421
14. Saaty TL (2010) Principia Mathematica Decernendi: mathematical principles of decision making. RWS Publications, Pittsburgh
15. Yoon K (1987) A reconciliation among discrete compromise situations. J Oper Res Soc 38(3):277–286
16. Yoon KP, Hwang C (1995) Multiple attribute decision making: an introduction. SAGE Publications, California

Index

Printed in the United States
By Bookmasters